EARTHQUAKE AND ATMOSPHERIC HAZARDS

Preparedness Studies

Edited by

M. I. EL-SABH

University of Quebec at Rimouski, Canada

S. VENKATESH

Environment Canada, Downsview, Canada

C. LOMNITZ

UNAM, Mexico City, Mexico

and

T. S. MURTY

Baird and Associates, Ottawa, Canada

Reprinted from Natural Hazards,
Volume 16, Nos. 2 & 3, 1997

SPRINGER SCIENCE+BUSINESS MEDIA, B.V.

A C.I.P. Catalogue record for this book is available from the Library of Congress.

ISBN 978-94-010-6113-1 ISBN 978-94-011-5034-7 (eBook)
DOI 10.1007/978-94-011-5034-7

Printed on acid-free paper

Table of Contents

Natural Hazards **16:** 101–102, 1997.

Preface

This issue of *Natural Hazards* contains a selection of papers that were presented at the Sixth International Symposium on Natural and Man-Made Hazards (HAZARDS-96) held in Toronto, Canada during 21–26 July 1996. The symposium was organized by the International Society for the Prevention and Mitigation of Natural Hazards (NHS), the IAPSO Commission on Natural Marine Hazards and the Tsunami Society. The symposium theme was 'Major Natural Disasters in the 90's – What can we learn from them?'.

HAZARDS-96 was very timely, given the large number of natural and man-made disasters that have already occurred in various parts of the world in the 1990's, the United Nations International Decade for Natural Disaster Reduction. The list is long and includes among others, volcanoes in the Philippines, storm surge-related flooding in Bangladesh, oil spills and fires resulting from the Middle East conflict, floods in North America and Europe caused by heavy rains, earthquakes and Tsunamis in Japan, hurricane damage in North and Central America. The human reaction and response to these disasters has varied widely from one disaster to the next and the economic and social costs have been immense with damages in the billions of dollars and significant loss of life.

With Japan's Kobe earthquake of 1995 fresh in everyone's mind, the symposium attracted a number of papers on that and other earthquakes. Several sessions during the symposium dealt with various aspects of this event. The symposium also included sessions on geological, meteorological and hydrological hazards, technological and man-made hazards, disaster mitigation and management, and insurance aspects of disasters. With participants from 18 countries around the world, over 120 papers were presented during the 30 sessions of the symposium. The abstracts of all the papers presented can be found in the symposium Abstracts Volume. Twenty complete manuscripts from the papers presented at the symposium were submitted for consideration in a post-symposium publication. They were subjected to rigorous scientific review and only 11 of these were accepted for publication in this special issue of *Natural Hazards* with emphasis on the preparedness aspects. A summary report on the HAZARDS-96 symposium is also included in this special issue.

We wish to thank the many individuals who offered their time and expertise to review the manuscripts and thus contributed to the technical quality of this volume. We also wish to acknowledge the support from the different organizations that cosponsored the symposium: the United Nations International Decade for Natural Disaster Reduction (IDNDR), Intergovernmental Oceanographic Commission

(IOC) of UNESCO, International Union for Geodesy and Geophysics, Environment Canada, Emergency Preparedness Canada, Canadian National Committee for IDNDR, Reinsurance Research Council (Canada), Emergency Measures Ontario and The Institute for Environmental Studies – University of Toronto.

M. I. El-Sabh
S. Venkatesh
C. Lomnitz
T. S. Murty

Natural Hazards **16:** 103–112, 1997.

Complexity Theory of Natural Disasters; Boundaries of Self-Structured Domains

A. E. SCHEIDEGGER
Technical University, Section of Geophysics, A-1040 Vienna, Austria

(Recieved: 16 October 1996; in final form: 14 August 1997)

Abstract. Disasters are often represented as complete breakdowns of quasi-stationary states in a landscape, but may also be part of the normal evolution of such states. A landscape is, in fact, an open, nonlinear, dynamic system where the tectonic uplift and the seismic activity represent the input, the mass wastage and the relief degradation the output. The apparent 'stability' is due to the fact that open, nonlinear dynamic systems tend to develop into relatively stable, self-organized ordered states 'at the edge of chaos', with a fractal attractor. Short of complete breakdown, such systems re-establish order in steps of various magnitudes which have a power-law distribution. Because of the fractal structure of the basic attractor, all subsets follow a power law which accounts for the distribution of the steps of recovery. As the domains of quasi-stationarity at the edge of chaos are represented by finite windows, the power-law does not cover all magnitudes. The stationarity windows are not only limited in range, but also in space and time. This should be taken into account in the assessment of hazards. Examples are given from seismology (earthquake frequency), volcanology (eruption frequency), river hydrology (flood frequency) and geomorphology (landslides).

Key words: complexity, chaos, hazards, limitations of attractors, seismicity, volcanism, river floods, landslides

1. Introduction

Very few simple fundamental principles govern the phenomenological development of a landscape. These are the *Principle of Antagonism*, the *Principle of Instability*, the *Catena Principle*, the *Selection Principle* and the *Principle of Tectonic Predesign* (Scheidegger, 1987). Amongst these, the Antagonism Principle is the most important. It states that there are two types of processes active in the formation of a landscape at any time: the endogenic (tectonic) and the exogenic (meteoric) processes. Generally, these two processes more or less balance each other so that a landscape represents the outcome of a steady-state process in an open, complex, nonlinear system. If any external parameter is changed, the steady state is usually re-established by means of a corresponding change of the other parameters (process-response theory). This change, even in a quasi-stationary state, may not be entirely continuous, but may occur in steps of various magnitudes. Thus, for 'disasters' to arise, it is not necessary that the system become completely unstable so that small random perturbations send it into complete chaos (regarding the fundamentals of 'chaos theory' cf., e.g., Cambel's [1993] book): 'disasters' are often represented by a stepwise recovery quasi-stationarity on the original fractal attractor owing to the attainment of stability limits in individual elements of the system. These steps represent a subset of the system attractor and are governed by a power law. In fact, a stationary landscape state does not correspond to dynamic *equilibrium*, but rather to *self-organized order at the edge of chaos*, including recovery, in an open nonlinear complex system.

2. Self-Ordered States

Geological 'disasters' are events in the evolution of landscape systems when the latter no longer follow a regular process-response pattern but leave a quasi-stable stationary attractor state. For this, it is necessary to analyze first the conditions producing stable states. There are innumerable examples in Nature of relatively stable self-organized ordered states at the edge of chaos (Bak *et al.*, 1988; Kauffman, 1993). The characteristic observables in such quasi-stationary states, such as the distributions of heights in a landscape, of mass in the Universe or in a sand pile generally, are spatially and temporally scale-invariant: they have been found to be fractal (Bak *et al.*, 1988). In a fractal set of dimension D, there exists a power law for subsets: The number N of subsets of (linear) 'size' L is proportional to $L\ \exp(-D)$. In Monte-Carlo sampling, this number N becomes the probability of encountering an 'event' of 'size' L. Such power laws have indeed been found in many complex natural systems. The frequent establishment of self-organized critical states in Nature (and, indeed, also in anthropogenic [such as social or economic]) systems strongly suggests the existence of a corresponding Universal Law of Nature (Bak *et al.*, 1987):

> Nonlinear open complex systems develop quasi-stationary states at the edge of chaos whose temporal signature is flicker noise and whose spatial signature is fractality with a dimension vastly lower than that of the phase space.

However, for a statement to be recognized and accepted as a 'Law of Nature', it is, according to Cohen and Stewart (1994), not enough that it is true; one has to *know* that it is true and, moreover, one has to be able to explain *why* it is true. Early studies on this problem have been made by Kolmogorov (1941) in connection with the theory of fully developed isotropic turbulence; Prigogine (1947, 1967; see also Nicolis and Prigogine, 1977) was the first to investigate the genesis of order on the edge of chaos ('ordered' or self-structured states have an entropy less than the maximum-possible entropy); Mandelbrot (e.g. 1983) applied the notion of fractals introduced by Hausdorff (1919) to various natural phenomena, and Turcotte (1992) summarized their usefulness in geology. However, the *necessity* for quasi-stationary states having to have a fractal nature seems to have been shown for the first time by Scheidegger (1995, 1996); we give here a brief description of the argument.

One can, first of all, look for conditions that are *necessary* for the spontaneous establishment of a quasi-stationary, ordered state. Most systems under consideration (e.g. landscapes) are dynamic systems (cf. the *Principle of Antagonism* explained in the Introduction): Elements are 'born' and 'die'. Evidently, for a stationary state to develop, the death rate must be equal to the growth rate in the system, otherwise one has complete obliteration or an explosion of the system.

Furthermore, all 'evolutionary' systems must be open and dissipative. Only in such systems do the laws of equlibrium thermodynamics not hold: the processes that occur in such systems are fundamentally irreversible and thus the thermodynamic theory of irreversible processes has to be applied: the usual form of the second law of thermodynamics does not apply. Prigogine (1947) noted that in open systems the entropy may well decrease during the approach to a steady state, i.e. its value at the nonequilibrium steady state ('ordered' or 'self-structured' state) may well be smaller than at equilibrium (death). In linear theory, a theorem of minimum entropy production holds for stationary states. Any (small) perturbation in such a stationary state always regresses; the steady states near equilibrium are essentially uniform in space if permitted by the external constraints. The stability of these states implies that *in a system obeying linear laws, the spontaneous emergence of order in the form of spatial or temporal patterns differing qualitatively from equilibrium-like behavior (death), is ruled out.* Moreover, any other type of order imposed on the system through the initial conditions is destroyed in the course of the evolution to the steady state (Nicolis and Prigogine, 1977, p. 46). Thus, a necessary condition for the spontaneous development of order is that the relations between the elements of the system under consideration *be nonlinear.*

An attempt at formulating a fundamental condition *causing* the establishment of a self-generated order in a complex system is represented by the statement of the *slaving principle* by Haken (1975), (also Haken and Wunderlin, 1988). This principle states that in the neighborhood of critical points, the behavior of a complex system is completely governed by few collective modes, the so-called order parameters, which slave all the other modes. In effect, the slaving principle of Haken has been shown to be equivalent to the statement that for a temporary order to be established, the corresponding (strange) attractor must be of a dimension which is vastly smaller than the dimension of the entire phase space: this prevents the system from wandering all over the entire phase space (= complete chaos) and confines it to a small region thereof (= 'relative' order or 'low-dimensional chaos'; cf. Kauffman (1993), pp. 178–179). In fact, the very concept of 'order' (stationary state with nonmaximum entropy) requires that of all the possible states of a system, only a few are permitted.

Next, one can ask why a stationary state at the edge of chaos has to be *fractal.* We have seen above that 'order', by the very definition of the concept, requires the corresponding attractor to have a dimension which is much smaller than the dimension of the entire phase space. However, there is no condition that this dimension be an integer: if it is not an integer, it is fractal – there are no other possibilities. Since the set of noninteger numbers is *vastly greater than that of integers*, the dimension of the attractor will generally be fractal.

The order at the edge of chaos has to be (quasi-)stationary (at least for a while; – otherwise the ordered state cannot be 'seen') with regard to small changes in the initial conditions. Thus, it is seen that ordered states are the *only* ones that have duration: there is a *'Darwinian selection' of ordered states*, much as the statically stable states are selected in a landscape according to the selection principle of Gerber (1969). Quite generally, in a system only those states can be 'seen' which have some duration. This is also the explanation of why evolution of every kind runs through 'punctuated' equilibrium stages (Gould and Eldredge, 1993): terrasses are formed in landscapes, biological species may develop in a short time-span and then remain stable for hundreds of millions of years (cf. primitive insects). Thus, the problem of explaining *why* order develops out of chaos is solved; the new natural law formulated above can be considered as *established* according to the criteria of Cohen and Stewart (1994).

Stationary states exist in a limited range of space and time only: It is well known that landscapes are not unifractal (Evans and McClean, 1995); the same can be said of sequences of disasters, such as earthquake series (Speidel and Mattson, 1993). Evidently, the domain of stationarity is only a *window* in phase space: its size must be large enough so that it is possible to observe and recognize the corresponding state as (quasi-)stationary. Everywhere in nonlinear systems, there are saturation effects: self-similarity cannot reach from minus to plus infinity. Often, the assessment of the size of this window is of fundamental importance in hazard assessment.

3. Examples

3.1. GENERAL REMARKS

The basic argument of this paper is that the (temporal) statistics of natural disaster phenomena can be understood as events occurring on a self-organized attractor at the edge of chaos in an open, nonlinear, complex system. In this instance, the term 'disaster' refers to any sudden deviation from a regular process-response curve – what one calls a 'disaster' is in any case only an anthropocentric valuation: if there are many *human* casualties, one speaks of a disaster, otherwise merely of a natural event (cf. the impact of the Tunguska meteorite in 1908, which would have been a catastrophic disaster if the object had hit New York). According to the general tenets of complexity theory (e.g. Cambel, 1993), such attractors have a fractal dimension which leads to 'power relationships' in the statistics of these events. However, such attractors are only *quasi*-stable, which means that they have the above structure in a limited range of time and space only. This leads to the *boundaries* of the validity of the predictive relationships. We shall give some specific examples from tectonics, volcanology, hydrology and geomorphology.

3.2. EARTHQUAKE STATISTICS

We take an earthquake series as the first example for a sequence of 'disasters'. An earthquake series in a country contains many events, small and big, only some of

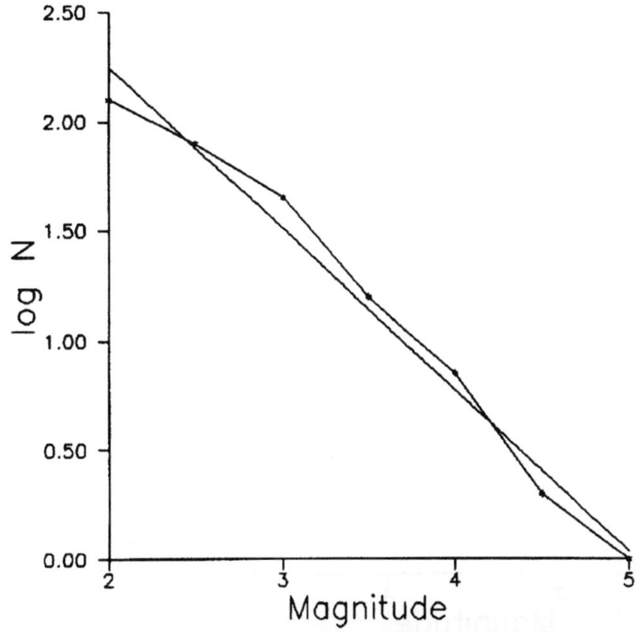

Figure 1. Gutenberg–Richter curve for the 126 earthquakes of the years 1971–1980 in Austria with magnitude $M > 2$, calculated from data published by Drimmel and Trapp (1982).

which could be called 'disasters'. However, for a theoretical discussion, *all* events have to be considered. Like in many other natural phenomena, the distribution of energy of earthquakes has been found to follow *grosso modo* a power law. Thus, the number N of earthquakes of magnitude M follows the Gutenberg–Richter (1949) law (a and b are constants; a depends on the time interval considered; b is a 'tectonic' parameter)

$$\log N(M) = a - bM.$$

Written in terms of the energy E ($M = c \log E$) this is for N(E) a 'power-law'. As a case in point, Figure 1 shows the magnitude-log N curve for the 126 earthquakes with $M > 2$ that occurred in Austria during the years 1971–1980, constructed from data published by Drimmel and Trapp (1982) – together with the best-fitting straight line, i.e. the 'Gutenberg–Richter' – curve for this series of earthquakes.

The fact that such Gutenberg–Richter curves can be constructed for many areas and time spans, has prompted scientists to interpret earthquakes, like many other natural phenomena in the geosciences, as examples of self-organized criticality (Bak and Tang, 1989; Turcotte, 1992). However, it should be noted that Gutenberg and Richter (1949) themselves had never claimed that their straight-line curve would reach from minus infinity to plus infinity: in every region they assumed the existence of a 'maximum possible magnitude' and, hence, their curve, of necessity, must eventually become vertical ('saturation effect'). Similarly, observations have been made by Speidel and Mattson (1993, 1995) indicating that the seismicity

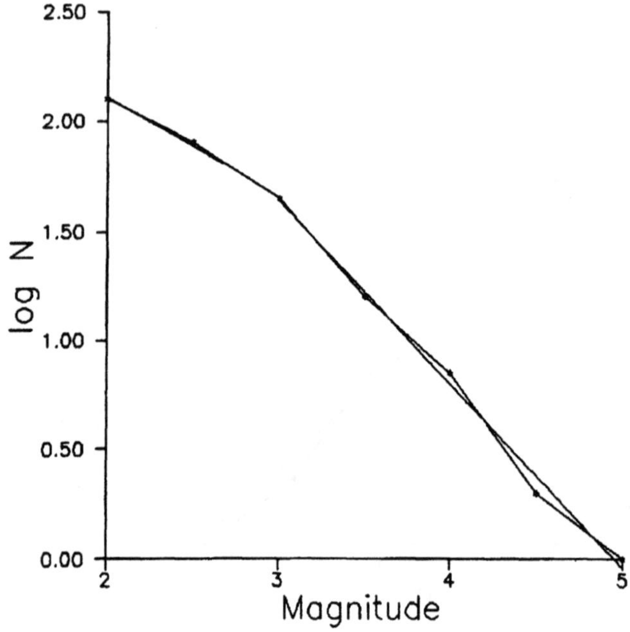

Figure 2. Data of Figure 1 fitted by two straight lines, one for magnitudes between 2 and 3 and the second for magnitudes between 3 and 5.

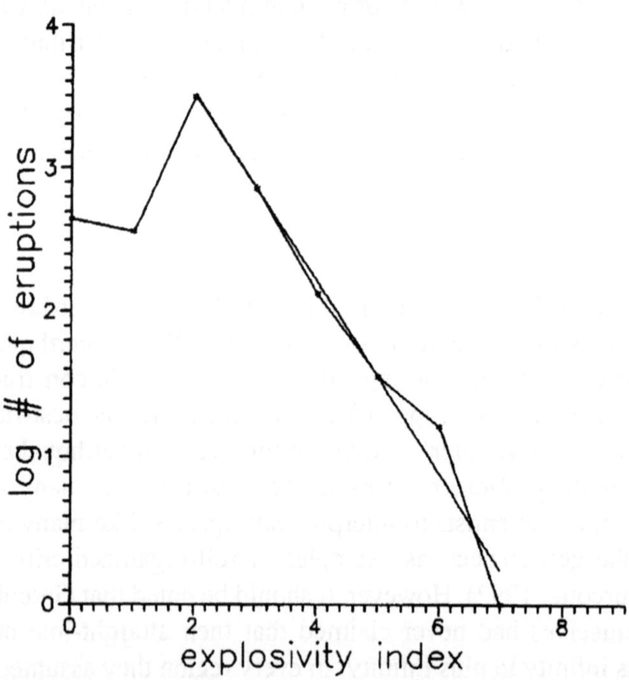

Figure 3. Plot of log # of volcanic eruptions in the world between 8000 BC and 1980 *versus* the explosivity index showing a straight 'Gutenberg–Richter line' between VEI 2 and 7.

statistics for any one area (such as the Central Mississippi Valley) might have to be considered as 'polymodal' rather than as corresponding to a single straight Gutenberg–Richter curve. An inspection of the data points in our Figure 1 also shows that the 'Gutenberg–Richter curve' should, in fact, have a kink near $M = 3$. We have therefore fitted *two* straight lines to the data, one for *low* magnitudes ($2 \leqslant M \leqslant 3$) and another for *high* magnitudes ($3 \leqslant M \leqslant 5$); the result is shown in Figure 2. For low and high magnitudes (there could also have been a detection problem for the low-magnitude earthquakes), one finds different best-fitting straight lines. There is, thus, a substantial difference in the statistical character of the two ranges. In addition, of course, there is the normal presence of a 'maximum possible magnitude'. This exemplifies the limited range of the self-structured states in a complex system.

3.3. VOLCANICITY

A similar case occurs in volcanological statistics: The volcanic eruptions of the world also show the basic fractality on a self-structured attractor at the edge of chaos, but again within a *limited* range only. Simkin *et al.* (1981) have made a catalog of the 4814 (identifiable) volcanic eruptions that occurred between 8000 BC and AD 1980. These were grouped according a '(Volcanic) Eruptivity Index' (VEI; – a dimensionless 'class' – identifier), defined by Simkin *et al.* (1981), which essentially corresponds to the logarithm of the volume of ejecta. It thus represents a kind of 'magnitude'. From the data of Simkin *et al.* (1981), by taking the logarithm of the numbers N of eruptions in the various VEI classes during the indicated period, a plot can be obtained (Figure 3). One sees that there is practically a straight course of the line from VEI = 2 to VEI = 7; the best-fitting straight line has been calculated and drawn for this range. Its existence indicates compliance with a fractal power law of the form

$$\log N(VEI) = a - bM$$

(a and b are constants). For VEI < 2 the curve drops off – this may have something to do with the poor record of small eruptions in (pre-)historic times; there is also a drop-off for VEI > 7 to $-\infty$ (not shown on the plot). Thus the validity of the normal 'order-law' is confined to the range $2 <$ VEI < 7, again showing the limitations of the latter.

3.4. STATISTICS OF RIVER FLOODS

A similar case can be made for river floods. Scheidegger (1975, p. 184) has collected the time series for the Gail river floods between 1890 and 1952. From this data, the number # of peaks exceeding a flow of Q (cu.m/sec) in the mentioned period is counted and then one can draw a plot of log Q (this corresponds to a 'magnitude' class M [dimensionless]) against log N (Figure 4), in which one sees practically a

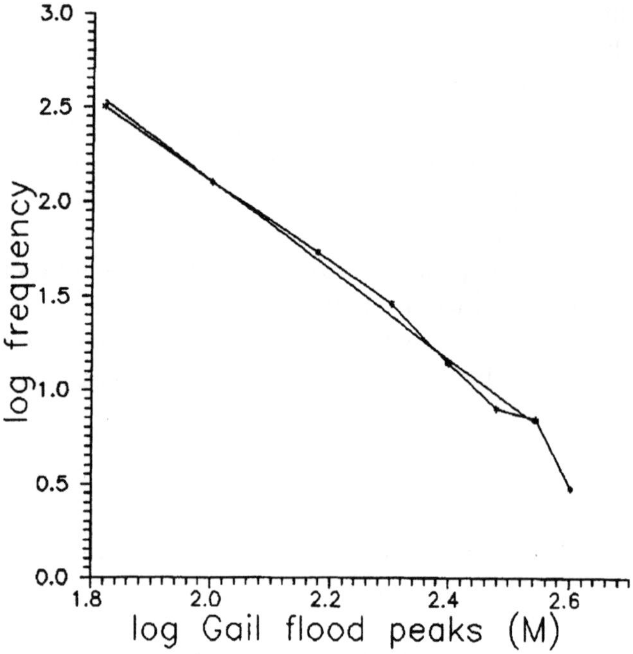

Figure 4. Plot of the logarithm of the number (frequency) of river flood (peaks) between 1890 and 1952 of the Gail River (in Carinthia, Austria) with a 'magnitude' (log Q [peak flow in cu. m]) > M *versus* that magnitude: again, a 'Gutenberg–Richter line' is obtained for $1.8 < M < 2.55$, showing a limited range of self-structured statistical stability.

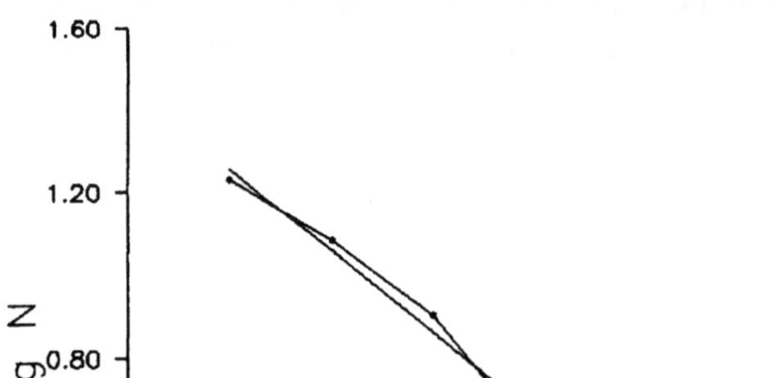

Figure 5. Plot of slide-magnitude M against log-frequency of occurrence in a particular valley in the mountains of Yunnan. Again, a *limited* range ($5 < M < 20$) for the validity of a 'Gutenberg–Richter curve' is found, indicating the boundaries of the corresponding self-structured domain in the complex system.

straight course between $M = 1.8$ and $M = 2.55$ (the plot shows the corresponding best-fitting line), indicating compliance with the fractal power law and therewith the existence of a self-structured domain; the last 'leg' of the plot drops off sharply, indicating the limitations of that law and therewith the boundaries of that domain.

3.5. LANDSLIDES

In a study of the Xiaojiang Basin, which lies near Kunming (NE Yunnan, China), Li and Wang (1984) have published a summary of the mud-bursts per year for 17 years in a particular ravine of that basin: the Jiangjia ravine. These mudflows are engendered by slides along the steep slopes of the ravine; thus one can take the number of outbursts in any one year as an integrated magnitude M of sliding in the ravine for that year. Then, it is possible to count the number N of years in which $m > M$ and to draw the plot of $\log N$ against M (Figure 5): one obtains a curve that corresponds to a fractal power law (for M vs. N; i.e. a straight line for M vs. $\log N$), but only between the magnitudes 4 and 20; the figure shows the best-fitting straight line between those limits. This shows that the land/mud slides in the mountain region investigated do indeed follow a power-law distribution in the main range; but for magnitudes below 5 and above 20 the power (respectively, linear in the logarithms) relationship is no longer maintained. This, again, shows the limitations of the fundamental relationship for self-structured systems.

4. Conclusions

In summary, we may reiterate that the explanation of many natural processes as representing the evolution of quasi-stationary self-ordered states at the edge of chaos in complex, open, nonlinear systems has been successful, at least for certain specific applications. However, the limitation of the domains ('windows') which such self-ordered states occupy in space and time, must be taken into account: the entire pattern of a quasi-stationary state may not only change with the range of parameters (such as magnitudes of earthquakes) considered, but also with the duration of the system.

References

Bak, P. and Tang, C.: 1989, Earthquakes as self-organized critical phenomenon, *J. Geophys. Res.* **94**, 15635–15637.

Bak, P., Tang, C., and Wiesenfeld, K.: 1988, Self-organized criticality, *Phys. Rev. A* **38**, 364–374.

Cambel, A. B.: 1993, *Applied Chaos Theory – A Paradigm for Complexity*, Academic Press, New York.

Cohen, J. and Stewart, I.: 1994, *The Collapse of Chaos*, Viking Press, New York.

Drimmel, J. and Trapp, E.: 1982, Die Erdbeben Oesterreichs 1971–1980, *Sitz.-ber. Oesterr. Akademie Wiss., Math.-natw. Kl., Abt.I*, **191**(1–4), 73–102.

Evans, I. S. and McClean, C. J.: 1995, The land surface is not unifractal: Variograms, cirque scale and allometry. *Z. Geomorphl. Suppl.* **101**, 127–147.

Gerber, E. K.: 1969, Bildung von Gratgipfeln und Felswaenden in den Alpen, *Z. Geomorph. Suppl.* **8**, 94–118.

Gould, S. J. and Eldredge, N.: 1993, Punctuated equilibrium comes of age, *Nature* **366**, 233–227.

Gutenberg, B. and Richter, C. F.: 1949, *Seismicity of the Earth and Associated Phenomena*, Princeton University Press, Princeton.

Haken, H.: 1975, Generalized Ginzburg-Landau equations for phase transition-like phenomena in lasers, nonlinear optics, hydrodynamics and chemical reactions, *Z. Physik B* **21**, 105–114.

Haken, H. and Wunderlin, A.: 1988, The slaving principle of synergetics, In: S., Lundqvist, N. H. March, and M. Tosi, (eds), *Order and Chaos in Nonlinear Physical Systems*, Plenum Press, New York, London, pp. 467–463.

Hausdorff, F.: 1919, Dimension und aeusseres Mass, *Math. Annalen* **79**, 157–179.

Kauffman, S. A.: 1993, *The Origins of Order: Self-Organization and Selection in Evolution*, Oxford University Press, Oxford.

Kolmogoroff, A.: 1941, The local structure of turbulence in incompressible viscous fluids for very large Reynolds numbers, *C.R. (Doklady) Aacad. Sci. USSR* **30**, 301–305.

Li, J. and Wang, J. R.: 1984, Mudflows in the Xiaojiang Basin, Yunnan Province, *Geooekodynamik* **5**(3), 143–258.

Mandelbrot, B.: 1983, *The Fractal Geometry of Nature*, Freeman, New York.

Nicolis, G. and Prigogine, I.: 1977, *Self-Organization in Non-Equilibrium Systems*, 8th printing, Wiley, New York.

Prigogine, I.: 1947, *Etude thermodynamique des phenomenes irreversibles*, Desoer, Liege.

Prigogine, I.: 1967, *Introduction to Thermodynamics of Irreversible Processes*, Wiley, New York.

Scheidegger, A. E.: 1975, *Physical Aspects of Natural Catastrophes*, Elsevier, Amsterdam.

Scheidegger, A. E.: 1987, The fundamental principles of landscape evolution, *Catena Supplement* **10**, 199–210.

Scheidegger, A. E.: 1995, Order at the edge of chaos in geophysics, *Abstracts IUGG XXI General Assembly* 1, A-11, UA51A-10.

Scheidegger, A. E.: 1996, Ordnung am Rande des Chaos: ein neues Naturgesetz, *Oesterr. Z. f. Vermesssung & Geoinformation* **84**(1), 69–74.

Simkin, T., Siebert,L., McClelland, L., Bridge,D., Newhall, C., and Latter, J. H.: 1981, *Volcanoes of the World*, Hutchinson Ross, Stroudsburg Pa.

Speidel, D. H. and Mattson, P. H.: 1993, The polymodal frequency-magnitude relationship of earthquakes, *Bull. Seismolog. Soc. America* **83**(6), 1893–1901.

Speidel, D. H. and Mattson, P. H.: 1995, Questions on the validity and utility of *b*-values: An example from the Central Mississippi Valley, *En Geol* **40**, 9–27.

Turcotte, D. L.: 1992, *Fractals and Chaos in Geology and Geophysics*, Cambridge University Press.

Natural Hazards **16:** 113–125, 1997.

March 1990 Hailstorm Damage in Sydney, Australia

KYLIE E. ANDREWS and RUSSELL J. BLONG
Natural Hazards Research Centre, Macquarie University, NSW 2109, Australia

(Recieved: 5 November 1996; in final form: 29 August 1997)

Abstract. On 18 March 1990, an intense hailstorm in Sydney, New South Wales, Australia caused insured damage valued at A$314 million – the third most expensive loss event in Australian insurance history.

While damage was widespread with claims for buildings spread across more than 130 postcodes, 20% of the claims came from just two postcodes. The proportion of dwellings of brick construction that made a claim was significantly less than the proportion of policies for this type of construction. Fibro (fibre-cement), timber and aluminium-clad dwellings are more likely to sustain damage than brick buildings in this type of storm.

Hail caused the majority of damage to exterior building items while water damage more commonly affected interior building items and contents. While the repair of Interior building items such as ceilings and walls cost more than contents damage on average, the repair/replacement costs of contents contributed more to the total costs. Damage to window glass and roofs together made up more than 50% of the total claim.

Key-words: hailstorm, hail damage, hail insurance losses, thunderstorms

1. Introduction

On 18 March 1990, an intense hailstorm struck Sydney, New South Wales, Australia, producing hailfalls in a swathe from southwest to northeast and causing extensive damage. Hailstones up to 8 cm in diameter were measured. Violent wind squalls and heavy rain were also reported. The distribution of maximum hailstone sizes by postcode (the Australian equivalent of a zip code) is shown in Figure 1, based on an extensive review of available sources. The largest hailstones were reported in and around the southwestern suburbs of Liverpool and nearby Bass Hill. Hail 5 cm in diameter or greater were also reported in Roseville and adjacent suburbs. A detailed analysis of hailfalls in Sydney since 1791 indicates that, in terms of maximum hailstone size, this event has an average recurrence interval of about once in 20 years; in terms of area affected the mean return period is about once in 25 years.

The storm formed at about 1:30 pm 105 km west-southwest of the Sydney CBD, and grew rapidly to thunderstorm intensity, reaching the southwestern outskirts of the city just after 3:00 pm. The most serious damage was sustained, between 3:40 and 4:00 pm when it passed over the Liverpool-Bass Hill-Yagoona area. Continuing to the east-northeast and moving at an average velocity of 44 km hr^{-1} (24 knots)

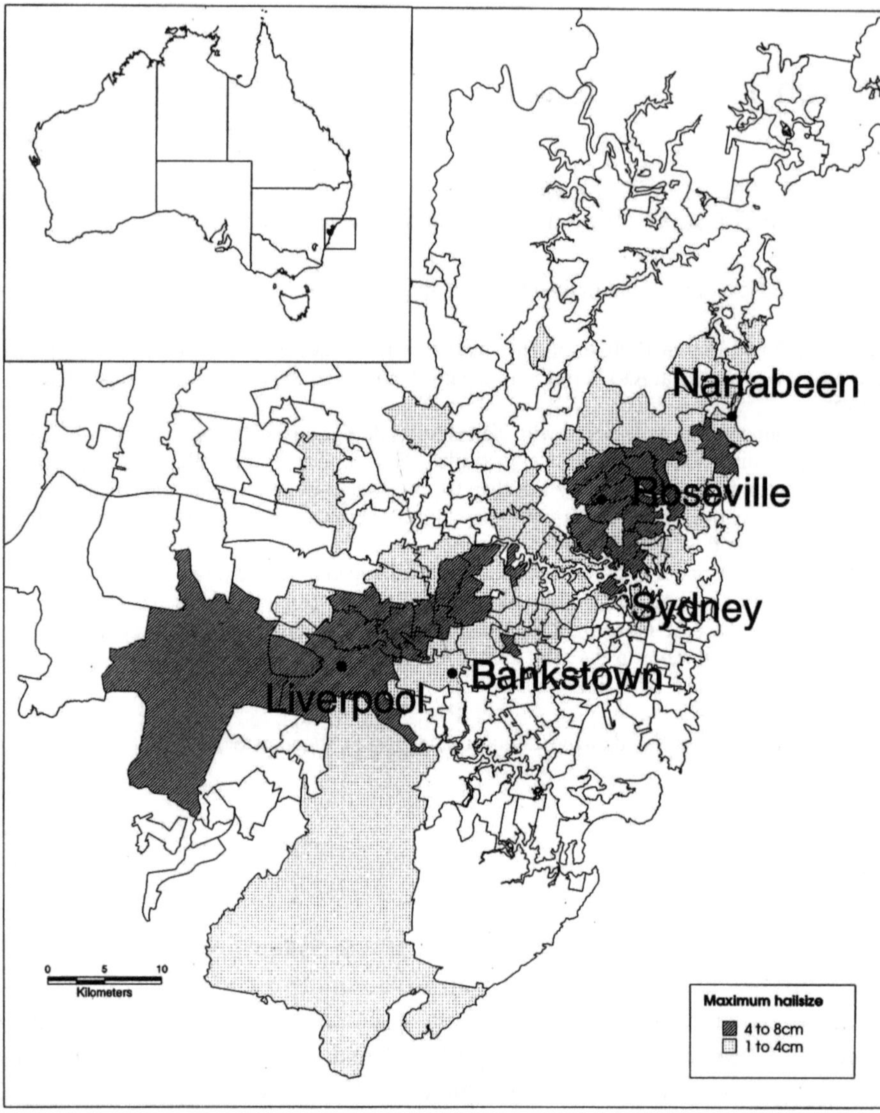

Figure 1. Maximum hailsize by postcodes in the 18 March 1990 hailstorm.

the storm crossed the coast near Narrabeen at approximately 4:40 pm (Mitchell and Griffiths, 1993).

The size of the hailstones produced by the 18 March 1990 storm indicate the presence of unusually strong updraughts (40 to 60 m s^{-1}) and the strength of the storm's out-flow winds imply that the downdrafts must also have been intense (Mitchell and Griffiths, 1993). There is some evidence that the hailstorm was a supercell thunderstorm (Mitchell and Griffiths, 1993).

Strong winds accompanying the storm unroofed houses and brought down trees and power lines. Seventy suburbs were blacked out and in some areas it took two days to restore power. At least 19 schools in the Liverpool area were seriously damaged, eight of which were temporarily closed. At one school, 267 windows were smashed (Mitchell and Griffiths, 1993). The anemometer at Bankstown Airport recorded a maximum wind gust of 109 km hr^{-1} (59 knots) at 3:54 pm, placing the storm in the class of a severe thunderstorm. The full impact of the storm was not felt by any other Bureau of Meteorology anemometers but damage indicates that the winds in the worst affected suburbs were considerably stronger than that recorded at Bankstown Airport (Mitchell and Griffiths, 1993).

Flash flooding also occurred in some areas with drains blocked by hail or wind-blown debris exacerbating problems. The heavy rainfall rates in the worst-affected areas for falls of 10 to 20 minutes duration had average return periods ranging from 20 to over 100 years (Mitchell and Griffiths, 1993). The heaviest rainfalls did not necessarily coincide with the areas experiencing maximum hailstone sizes.

The total insured damage bill from the 18 March 1990 event is estimated at A$314 million (about US$240 million). Insurance Council of Australia data (including that of some nonmember companies) indicate that at mid-December 1991 a total 84 354 claims had been paid. Of the total payment 35% was for domestic house claims, 17% for commercial properties and 43% for motor vehicles. The March 1990 hailstorm is the most damaging thunderstorm on record in Australia and the third largest loss event in Australian insurance history (after the Newcastle earthquake, 1989, and Cyclone Tracy, 1974). There were no deaths or reports of serious injury.

While research has been undertaken to assess the impact of hailstones when fired at different materials in a controlled environment (Armstrong *et al.*, 1980; Cox and Armstrong, 1979; Greenfeld, 1969; Jenkins and Mathey, 1982; US Dept of Energy, 1978; Vorobieff *et al.*, n.d.), we are unaware of comprehensive research analysing the cause and nature of damage to domestic buildings in real thunderstorms. The current research aims to:

1. Determine the distribution of damage caused by the 18 March 1990 hailstorm, in terms of the percentage of the total sum insured and percentage of policies affected.
2. Identify the domestic building and contents items most vulnerable to damage and to establish the costs of repair/replacement of these items.
3. Establish relationships between the items damaged, the costs of the damage and the causes of the damage.
4. Develop relationships between losses and hailstone sizes in order to establish percent loss curves for the March 1990 hailstorm.

2. Data and Analysis

In this study attention is focused on maximum hailstone size per postcode as a unit of measurement of the severity of the hailstorm from one location to the other. It is recognised that there is a natural tendency amongst observers to overestimate both hailstone size and damage produced, with some studies finding that on average, the public exaggerated the maximum size of hailstones by 20% (Charlton *et al.*, 1995). However, few attempts have been made to estimate mean hailstone size and it seems likely that damage to structures is more highly correlated with maximum hailstone size than it is to modal hail size, impact energy or number of hailstones per unit area (Summers and Wojtiw, 1971; Chagnon, 1977).

Data were obtained from several Australian insurance companies who provided information on claims made in relation to the March 1990 hailstorm. The quality of the data is varied, and claim amount, building construction type, year of construction, address of the insured property were supplied in the best, but not all, cases. In order to determine the proportion of damage to the portfolios, policy data at the time of the storm were supplied by some companies.

The analysis includes a total of 105 000 policies, with a Total Sum Insured (TSI) of A$12 billion. The maximum claim recorded was for $57 974. The claims represent 0.18% of the TSI. This figure can be compared with the 0.103% and 0.039% for buildings and contents respectively reported by Andrews *et al.* (1995) for the same event.

A sample of claims were chosen from the portfolio of one insurer who contributed to the portfolio analysis. The sample chosen for analysis represented 20% of claims made to that insurer in relation to the March 1990 hailstorm. The sample mimics the distribution of all claims made to that insurer both spatially (in terms of the number of claims for each postcode) and financially (in terms of the total cost of the claims). This sample is assumed to be representative of the total body of claims from that insurer.

Analysis of the sample of claims files resulted in a database providing information about (1) items damaged in the hailstorm; (2) the cost of the damage; and (3) the cause of damage.

Items damaged were divided into exterior building, interior building and contents. Exterior building, for example, includes walls, exterior cladding, roofs, garages and sheds, fences, window glass and skylights. Interior building elements include ceilings, cupboards, electrical wiring, floors, plaster and wallpaper. Contents items include bedding, blinds, books, furniture, TVs, VCRs and other electrical goods. These divisions were made after discussions with insurers. A general category was used when the elements damaged were either not specified or lumped together.

The most obvious cause of damage for each item damaged was listed in the database. Assigning causality was based on either a value judgement made by the

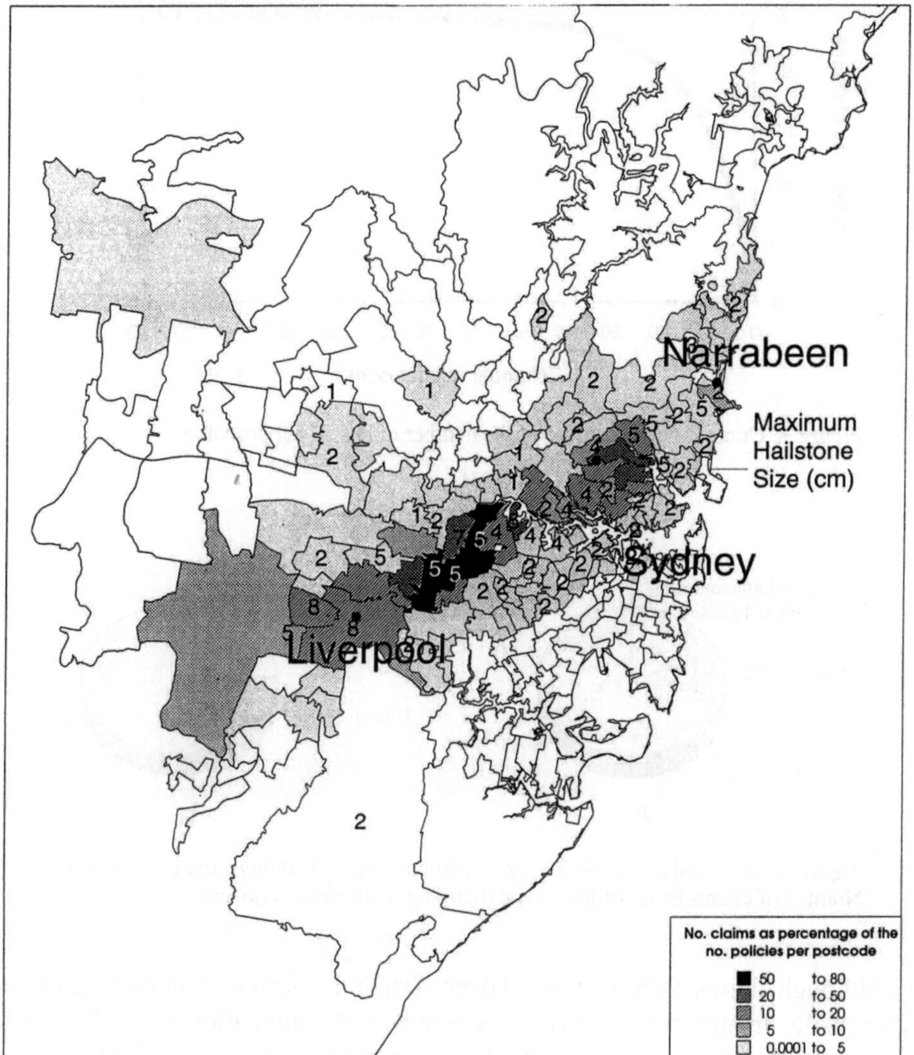

Figure 2. The number of claims made in relation to the 18 March 1990 storm in each postcode as a percentage of the number of policies in each postcode, with the maximum hailsize overlayed.

researcher entering information, or on comments within the claim file that described a particular cause as being responsible for the damage to an item.

3. Claims Distribution

The concentration of claims around the suburbs of Lidcombe and Liverpool is illustrated in Figure 2, which gives the number of claims as a percentage of the number of policies in a postcode. The pattern in this figure corresponds well with the path of the storm displayed in Figure 1. In the worst affected postcodes, up to 80% of policies made a claim for damage caused by the hailstorm.

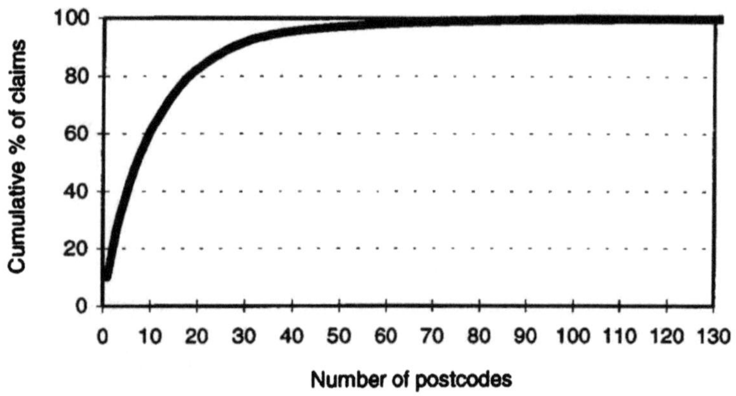

Figure 3. Cumulative percentage of the number of claims per postcode.

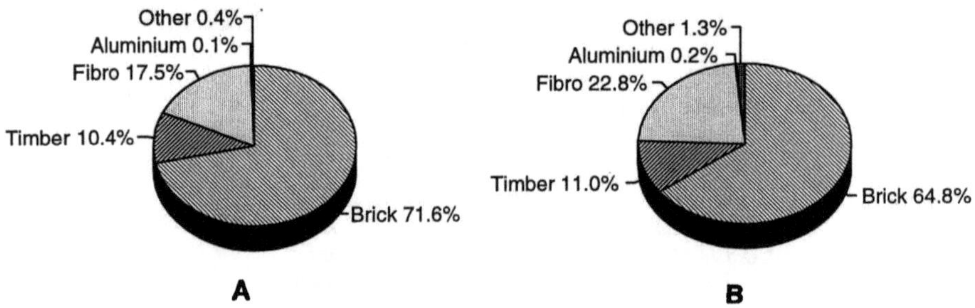

Figure 4. A: Number of policies by construction type (buildings and contents policies). B: Number of claims by constrution type (buildings and contents policies.

Although claims were distributed over 52% of the postcodes in the Sydney area (specifically, Insurance Council of Australia Risk Accumulation Zones 41–44), the majority of the claims occurred in a concentrated area (Figure 3). Eighty percent of claims occurred in only 18 postcodes and 20% of claims came from just two postcodes – Lidcombe and Liverpool.

4. Vulnerable Elements: Construction Type

Policies are divided into five construction types – brick (including brick veneer), timber, fibro (fibre-cement), aluminium and other (which includes stone, iron, cement block, etc). Comparing the breakdown of policies into building construction types with a similar breakdown for claims reveals that the proportion of brick dwellings making a claim is significantly less than the associated proportion in the policy data (Figure 4) – 64.8% and 71.6%, respectively. These data indicate that brick buildings were less vulnerable to damage in this storm than the other construction types.

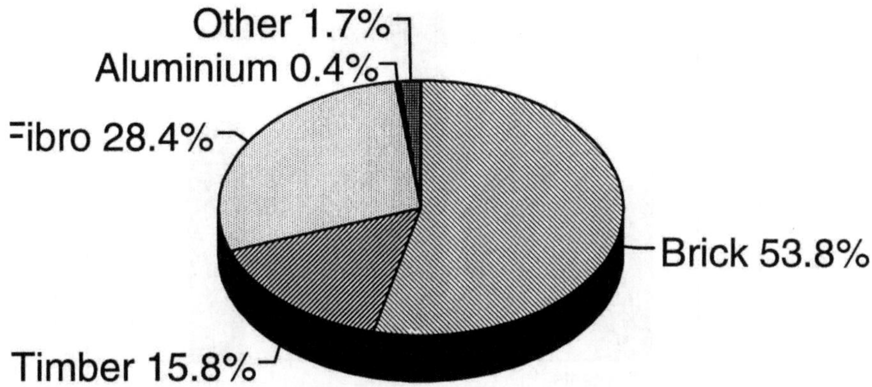

Figure 5. Sum of claims amounts by construction type (buildings and contents policies).

Figure 5 shows the sums of claims amounts by construction types. Taken together with Figure 4, these data indicate that while damage to brick buildings make up the majority of the insured loss from the storm, it is the dwellings constructed of fibro, timber and aluminium which are more likely to sustain damage.

5. Vulnerable Elements: Individual Items

Various studies indicate that tiles, gutters, windows, brittle cladding materials and metal sheeting are all at risk during heavy hail. Thin metal sheeting is dented or even penetrated, while tiles can develop hairline cracks. Age and impact location are important factors for many materials (Vorobieff *et al.*, n.d.). Asphalt shingle replacement seems to have reached more than 50% in large areas following the Calgary, 1991 storm (Charlton *et al.*, 1995), but we are unaware of detailed studies of the range of items actually damaged in a storm.

In the claims files examined a total of 1 364 items were mentioned as damaged. About 68% of these were Exterior building, nearly 11% Interior building and 21% Contents items.

Average replacement costs range from $870 for contents, through $1200 for interior buildings items to $1750 for Exterior building items (Figure 6). However, Figure 6 also reveals that while repair of interior building items cost more than contents damage on average, the repair/replacement of contents contributed more to the total cost.

6. Exterior Building

Of the exterior building items, damage to the exterior structure (roof and walls but excluding windows, doors, screens, etc.), of houses, garages and sheds made up 31% of elements while the nonstructural components (windows, doors, awnings,

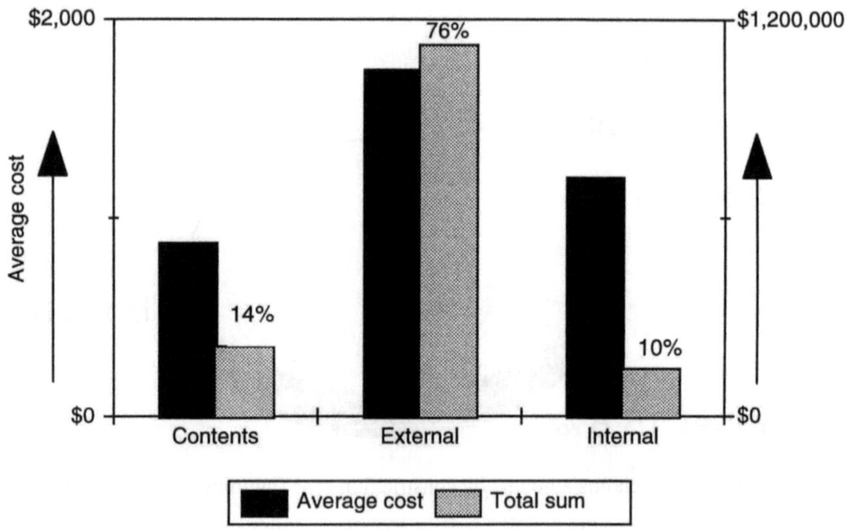

Figure 6. Average cost and total cost of items damaged by element type (the percentage of the total sum is also given).

screens, guttering, fascia, etc.) composed 64% of items. Miscellaneous exterior items such as fences, lights, pools and staircases total less than 5% of the damaged elements.

Damage to glass, including windows, lights, and solar panels, totalled one third of items for which claims were made. Roof tiles, roof cladding, guttering and TV antennae damage also amounted to one third.

Exterior walls and openings (windows, doors, flyscreens, etc.), made up 52.5% of the damaged items; as these elements are virtually all sheltered from vertically-falling hail and rain, it is clear that wind or wind-driven hail and rain were important agents.

Tiled roofs were, overwhelmingly, the most commonly damaged roof type – not surprisingly, given the dominance of clay tile roofs in Sydney. Unfortunately, we have no policy data specifying roof types, but we can say that 15% of the claims examined that noted the presence of a tiled roof made claims for damage to the roof. Roof materials can be ranked in decreasing order of susceptibility to damage: aluminium, fibro, tiles, iron. The average total cost of roof repairs (all types including Unknown) was about $1650. Roof repairs (including roof points) contributed 30% of the total cost of external repairs.

The average cost of repairs to clear window glass was about $2250, with a total cost of repairs representing 41% of all external repair costs. When repairs to other windows, lights, skylights and solar collectors plus window flyscreens, shutters and window frames are added to this total, the percentage of the total external repair costs exceeds 43%.

Claims for cladding contributed nearly 11% of the total external claims amounts. When fascias, other external walls and paintwork are added to this total, the proportion of the total external claim rises to over 13%. The total damage to guttering and downpipes in the sample amounted to only 2% of the external damage claim.

7. Interior Building

Within the interior structure of the house, damage to walls, ceilings (including insulation) and floors comprised 78% of damaged items. Ceilings and internal paintwork were the items damaged most frequently. Miscellaneous items such as shower screens, cupboards and mirrors, totalled less than 5% of the damaged items in the sample.

In terms of total costs, internal paintwork was the most expensive item, representing 38% of the total internal building repair costs. Ceiling damage was the next most expensive item at 31% while ceilings plus insulation batts amounted to 43% of the total and damage to internal walls and wallpaper amounted to 6% of the total. Presumably, a significant proportion of the paintwork costs also relate to internal walls. Damage to floors and floor coverings (excluding carpets) reached more than 6% of the total Internal repair/replacement cost.

8. Contents

Carpets and window coverings (blinds and curtains) together total 47% of the contents items reported damaged, and along with indoor furniture and electrical equipment were the most frequently damaged contents items. Indoors, tables, sofas and lounges seem to have been the most common categories of furniture damaged. Most of the claims for damage to floors or floor coverings were claims for damaged carpet.

About one tenth of the contents items reported damaged can be regarded as outdoor items – outdoor furniture, airconditioners, clothes lines, lawnmowers, pool liners, letter boxes etc.

Although the highest average cost of repair/replacement for contents items was for sofas/lounges, the total cost of all furniture items (indoor and outdoor) reached nearly 18% of the total Contents bill for the sample considered here. Carpets were the most commonly claimed item and the largest proportion of the total contents claim at more than 50%. Blind and curtain claims totalled nearly 14% while electrical equipment totalled more than 9% of the total contents claims amount.

9. Replacement Cost Overview

Table I provides an overview of the major items contributing to the total cost of the sampled damage. Damage to window glass and roofs together make up more than 50% of the total claims, while external cladding, floors and floor coverings

Table I. Overview of major cost items

Item	Location	% Total cost[a]
Roofs	External	22.8
Windows + glass + screens	External	33.1
Cladding + walls + paintwork	External	9.9
Guttering + downpipes	External	1.6
Internal paintwork	Internal	3.7
Ceilings + insulation	Internal	4.2
Walls + wallpaper	Internal	0.6
Floors + carpets + other coverings	Internal	8.0
Furniture (indoor + outdoor)	Contents	2.5
Blinds + curtains	Contents	2.0
Electrical equipment	Contents	1.4

[a] Items in this table do not sum to 100%.

contribute another 16%. Ceilings, insulation materials and internal paintwork contribute an additional 8%. The items listed above, together with associated damage, contributed more than 80% of the total damage bill for the sample examined. The cost of repairs/replacement to electrical equipment was less than that for blinds and curtains or indoor furniture. Similarly, damage to guttering and downpipes is a small percentage, though some such damage may be included under damage to roofs. The items listed in Table I together account for nearly 90% of the total damage bill.

10. Cause of Damage

Thunderstorms produce damage through many agents – wind, hail, rain, flood, or tree fall. Assigning damage to specific agents is not easy. However, in the records available to us, damage has often been ascribed to specific agents on claim forms and/or in loss assessors' reports.

Figure 7 summarises some of this information for two insurers. We cannot be certain that terminology has been used consistently by those reporting but the emphasis on hail damage and water damage is obvious. Surprisingly, wind damage is reported directly in only a few cases though tree fall is probably the result of wind and 'storm general' and 'storm/hail' categories may include some wind damage.

Hail is the most frequent cause of damage to exterior building items while water damage is most frequently the cause of damage to Interior building items and contents (Figure 8). Figure 8 emphasises the importance of both hail and water damage in the 18 March 1990 storm. Around 90% of all items – exterior building, interior building and contents – were damaged by either hail or water. 58% of building policies which reported hail damage also reported water damage. When

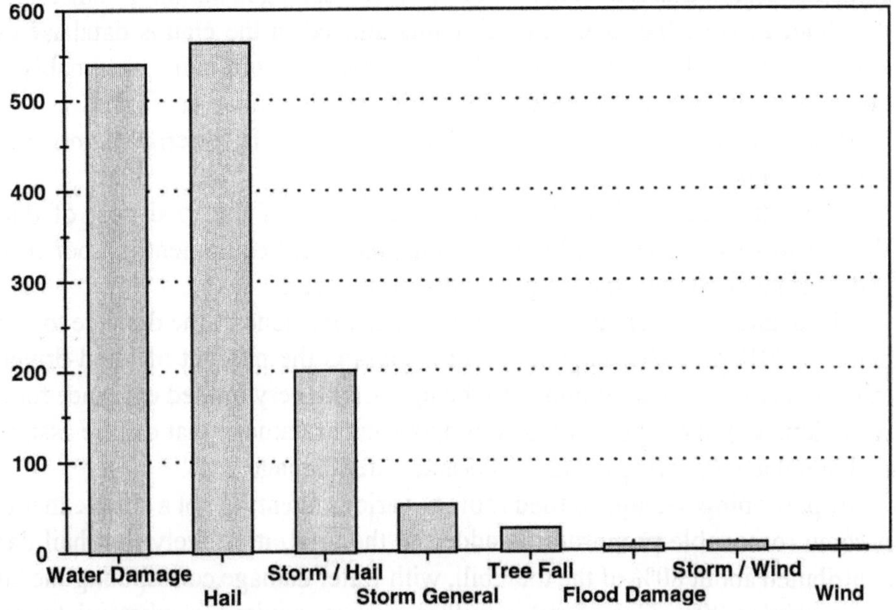

Figure 7. Cause of damage to items in the 18 March 1990 hailstorm. The vertical axis indicates the number of items damaged.

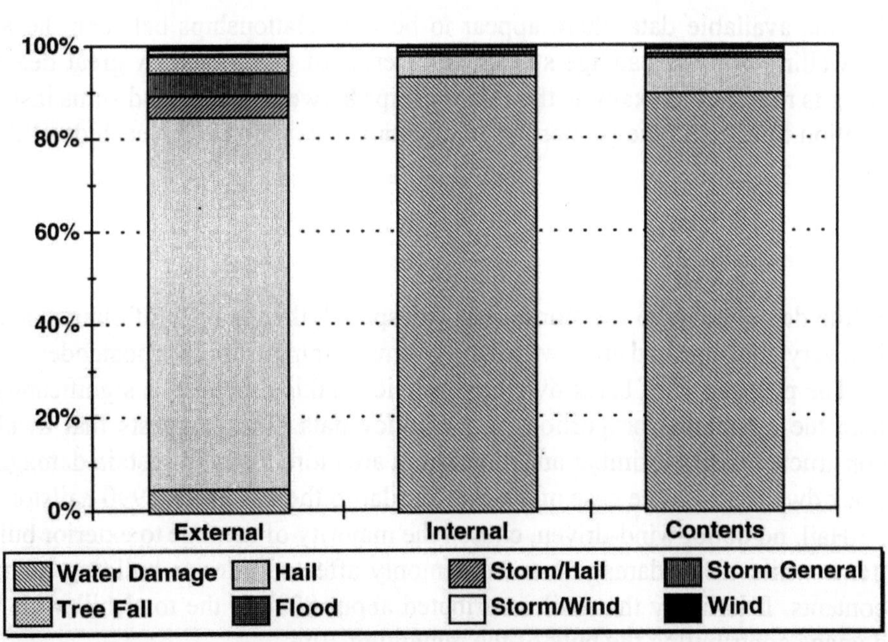

Figure 8. Cause of damage to external building, internal building and contents.

contents and building items were claimed together, in 80% of cases water-damaged contents were claimed in conjunction with hail-damaged building items.

About 27% of the total cost of claims utilised in the claims database can be assigned to hail damage. Most of this cost resulted from hail (presumably wind-driven hail) damage to aluminium cladding and aluminium roofs. Lesser amounts were contributed by fibre-cement and fibre glass roofs, external paintwork, and window frames.

Water damage resulted in approximately 73% of the total cost of damage. Water damaged carpets, ceilings, curtains, electrical equipment, indoor furniture and internal paintwork made up the bulk of this total.

The available information explicitly apportions none of the damage to window glass. If 90% of the damage to window glass is the product of wind-driven hail impact (a reasonable assumption in the light of the very limited evidence for direct wind damage), then the readjusted proportions of damage that can be assigned to individual agents swings to hail 74% and water damage 26%.

Apportioning damage to tiled roofs to various agents is not a simple matter, but if some reasonable proportion is added to the total, it is likely that hail damage contributed about 80% of the total bill, with water damage contributing the bulk of the remaining 20%. Considerable further work is required to substantiate some of these details.

11. Dwelling Age and Damage

For the available data, there appear to be few relationships between the age of a dwelling and the damage sustained in terms of claim cost. A great deal more work is required to examine the relationships between claims and sums insured in relation to age and the proportion of claims versus policies for each dwelling age group.

12. Conclusions

While damage due to the storm was widespread, the majority of claims occurred in a very concentrated area, with 20% claims coming from two postcodes.

The proportion of brick dwelling policies making a claim is significantly less than the associated proportion in the policy data. This suggests that dwellings constructed of fibro, timber and aluminium are more likely to sustain damage than brick dwellings, in the case of a storm similar to the 18 March 1990 hailstorm.

Hail, no doubt wind-driven, caused the majority of damage to exterior building items while water damage more commonly affected interior building items and contents. It is likely that hail contributed about 80% of the total bill with water damage contributing the bulk of the remaining 20%.

Damage to window glass and roofs together made up more than 50% of the total claims. While the repair of interior building items such as ceilings and walls cost more than contents damage on average, the repair/replacement of contents contributed more to the total cost.

This analysis has opened a window on to the constituents of storm damage, the agents of damage, and the items damaged, as well as providing an overview of the distribution of damage caused by the 18 March 1990 hailstorm. The review prompts several questions, primarily: Will analysis of other storms produce similar statistics on items damaged, costs, and causes?

Similar analyses on other storm events will aid in the development of our understanding of disaster events. An understanding of the way building age and other parameters affect damage could be invaluable. The possibility is that, with this understanding, we will be better equipped for the reduction of damage from storm events.

Acknowledgements

We are grateful to a number of direct insurers, reinsurers and reinsurance brokers who have financially supported the Natural Hazards Research Centre and the hailstorm project as well as providing valuable data on which much of this paper is based. The Australian Bureau of Meteorology also provided some financial support at an early stage of our hailstorm investigations. Clare Byrnes played an invaluable role in reading claims information and data analysis.

References

Andrews, T., Carless, R., Vaccaro, G., and Goodwin, M.: 1995, Catastrophe PML estimation, *Institute of Actuaries of Australia 1995 General Insurance Seminar*, 26–29 November, Launceston.

Armstrong P. R., Cox, M., and de Winter, F.: 1980, *Need for and Evaluation of Hail Protection Devices for Solar Flat Plate Collectors*, Atlas Corporation ALO-4291-2.

Changnon, S. A.: 1977, The scales of hail, *J. Appl. Meteorol.* **16**, 626–648.

Charlton, R. B., Kachman, B. M., and Wojtiw, L.: 1995, Urban hailstorms: a view from Alberta, *Natural Hazards* **12**, 29–75.

Cox. M. and Armstrong, P. R.: 1979, *A Statistical Model for Assessing the Risk of Hail Damage to any Ground Installation,* Atlas Corporation ALO-4291-1.

Greenfeld, S. H.: 1969, *Hail Resistance of Roofing Products,* Institute for Applied Technology, Building Research Division, Washington, DC.

Jenkins, D. R. and Mathey, R. G.: 1982, Hail impact testing procedure for solar collector covers, U.S. Department of Commerce NBSIR 82-2487.

Mitchell, E. and Griffiths, D.: 1993, Report on the Sydney hailstorm: March 1990, Bureau of Meteorology Report.

Summers, P. W. and Wojtiw, L.: 1971, The economic impact of hail damage in Alberta, Canada and its dependence on various hailfall parameters, *Seventh Conference on Severe Local Storms*, 5–7 Oct 1971, Kansas City, pp. 158–163.

United States Dept of Energy: 1978, Central receiver solar thermal power system – hailstone simulation test report, DOE EY-76-0-03-1111 (Modification No. A006).

Vorobieff, G., Paterson, D., and Knightley. R.: date unknown, Hailstone evaluation of cladding materials, CSIRO Division of Building Construction and Engineering (internal report).

Natural Hazards **16:** 127–134, 1997.

A Test of the Cyclicity of Earthquakes

G. R. DARGAHI-NOUBARY
Department of Mathematics, Bloomsburg University, Bloomsburg, PA 17815-1301, U.S.A.

(Recieved: 12 November 1996; in final form: 26 June 1997)

Abstract. The periodogram analysis of time intervals between successive earthquakes is used for testing the seismic gap hypothesis. Using the available data, it is shown that the existence of a periodic component can be rejected with high confidence. The test provides evidence against quasi-periodicity and regular clustering of the arrival times.

Key words: cyclicity of earthquakes, time series, periodogram.

1. Introduction

An earthquake catalog is a chronologically ordered sequence identified at least by magnitudes and origin times. For many years, there have been attempts to determine whether regularities in occurrence can be found in a catalog of earthquake events, in the hope that seismic history can be meaningfully extrapolated. See Lomnitz (1994, Ch. 7) for an excellent account of this and related subjects.

In order to translate the presumed regularities into the language of modeling, some assumptions are needed. It is often assumed that the stored elastic energy, or potential energy of deformation, of an earthquake fault system is increased randomly and nonmonotonically toward a maximum. At this point a seismic catastrophe occurs in which the maximum possible energy is released in an earthquake. The catastrophe suddenly reduces the potential energy stored in the system to a low value, and the process of increase of potential energy begins once again. Because of inhomogeneity along the fault surface, we can imagine that the mechanism of storing energy of deformation along the fault surface may have a stochastic character on a scale of times that are long compared with the intervals between catastrophes. The shocks are assumed to be instantaneous. Note that the potential energy is presumed to be derived from tectonic deformation of the blocks or plates abutting on one another in the neighborhood of the fault.

Ideas of this nature have led to what is known as the seismic gap hypothesis. This hypothesis states that earthquake hazard is small immediately following the previous large earthquake and increases with time since the last large event on certain fault or plate boundaries (see, e.g., Sykes and Nishenko, 1984, p. 591). Recently, Kagan and Jackson (1991) have tested and rejected the seismic gap hypothesis. They considered the times of large earthquakes as a point process and used a coefficient of variation, defined as the ratio of the standard deviation of interval times to the mean recurrence time. A quasi-periodic process will have a coefficient of variation less than one, while a clustering process will have a coefficient of variation greater than one. The Poisson process has a coefficient of variation equal to one and predicts a seismic hazard independent of the time of occurrence of the previous seismic event. Note that the seismic gap hypothesis assumes that earthquake occurrence is a quasi-periodic process, so that the earthquake potential is low when the elapsed time since the last large earthquake is less than the mean recurrence time and is high afterward. It is interesting to note that a paleoseismic record at Pallett Creek, California, is the longest reliable record and suggests a clustering process cannot be rejected (Sieh *et al.*, 1989).

Recently the seismic gap hypothesis has been used in long-term forecasting of earthquake around the whole Pacific rim (e.g. Nishenko, 1991), in California (Sykes and Niskenko, 1984; Bakun and Lindh, 1985; Working Group on California Earthquake Probabilities, 1988; Lomînitz, 1985, 1994), and in other Pacific regions (e.g. Kelleher *et al.*, 1973). Because of the scientific and social importance of these forecasts, the seismic gap hypothesis deserves rigorous testing.

The seismic gap hypothesis can be tested using different approaches. For example, it is possible to consider an appropriate bivariate exponential distribution to model the size and the time between earthquakes. On the one hand, the magnitude of an earthquake is often taken to have an exponential distribution leading to the well-known Gutenberg–Richter law. On the other hand, earthquakes are assumed to occur according to a Poisson process, implying that the time between two successive event is an exponential random variable. Expect for some simple cases an analysis of the simultaneous considering of these two variables using bivariate exponential distribution has never been performed, see, e.g., Epstein and Lomnitz (1966) and Lomnitz (1994). Here a zero correlation will lead to rejection of the seismic gap hypothesis. This approach is investigated in Dargahi-Noubry and Razzaghi (1994).

In this paper, an approach is introduced for testing the possibility of a periodic component. This is presented in the next section. Numerical results based on earthquakes in central U.S.A. and in circum-Pacific region will be used for demonstration.

2. Testing Procedure

In this section we describe a testing procedure based on the periodogram introduced by Schuster (1898) for detection of periodicities in time series. Suppose $(X_1, X_2, \ldots, X_n)$ is a sample from the time series $\{X_t\}$. Define the periodogram

$$I_N(\omega) = \frac{2}{N}\left|\sum_{t=1}^{N} X_t\, \mathrm{e}^{i\omega t}\right|^2 = A^2(\omega) + B^2(\omega),$$

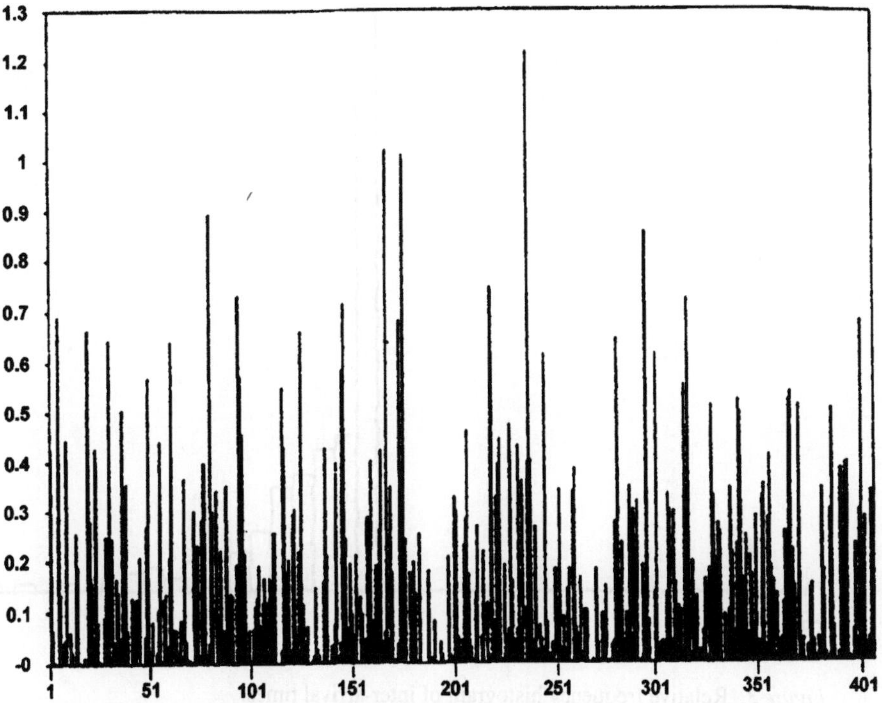

Figure 1. Time intervals between earthquakes since 1920 (days).

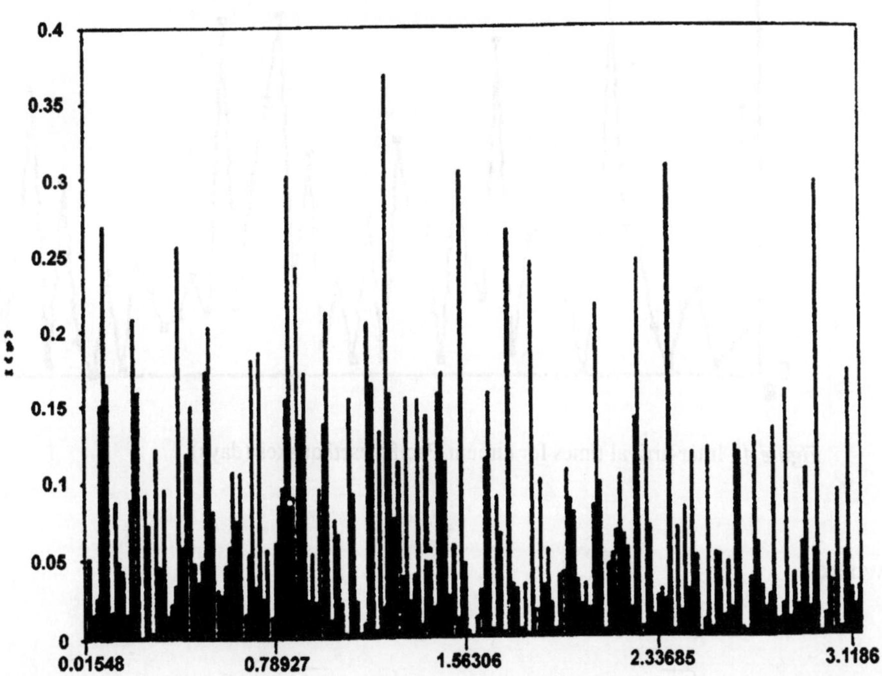

Figure 2. Normalized periodogram of inter-arrival times of earthquakes after 1920 (radian).

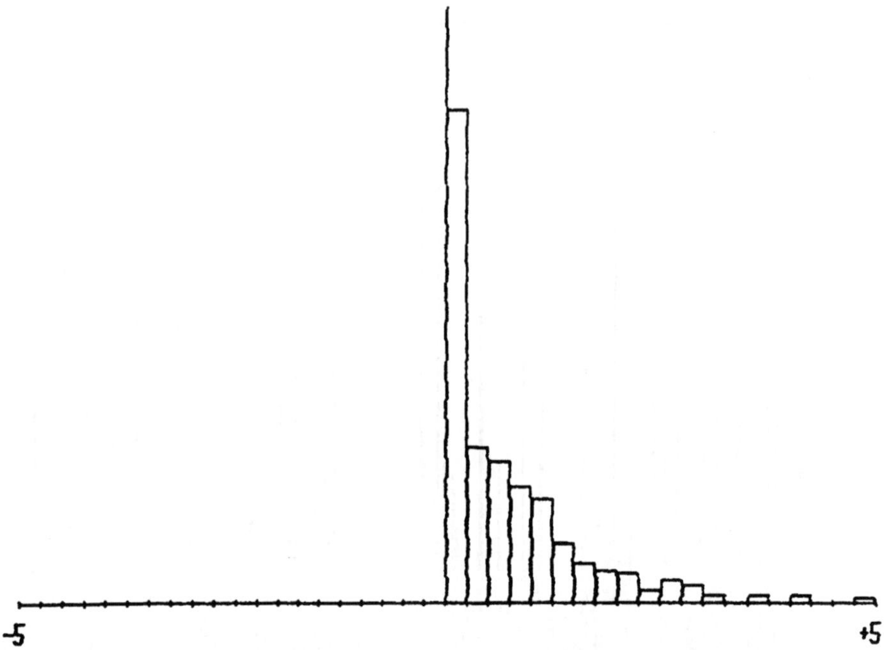

Figure 3. Relative frequency histogram of inter-arrival times.

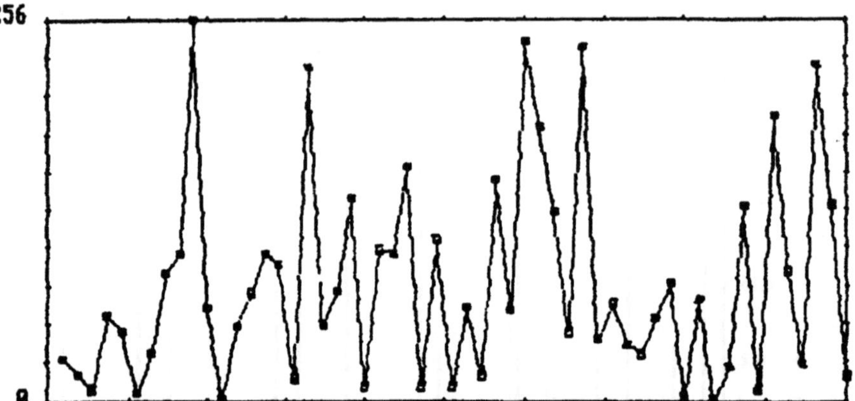

Figure 4. Inter-arrival times for circum-Pacific earthquakes (days).

where

$$A(\omega) = \sqrt{\frac{2}{N}} \sum X_t \cos \omega t, \quad B(\omega) = \sqrt{\frac{2}{N}} \sum X_t \sin \omega t.$$

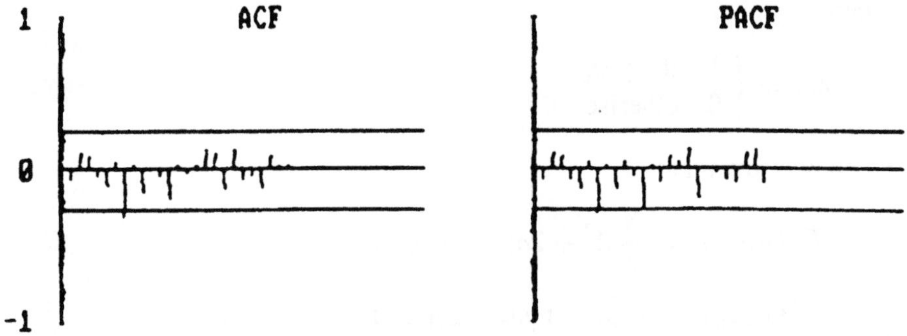

Figure 5. Autocorrelation and partial autocorrelation functions for data in Figure 4.

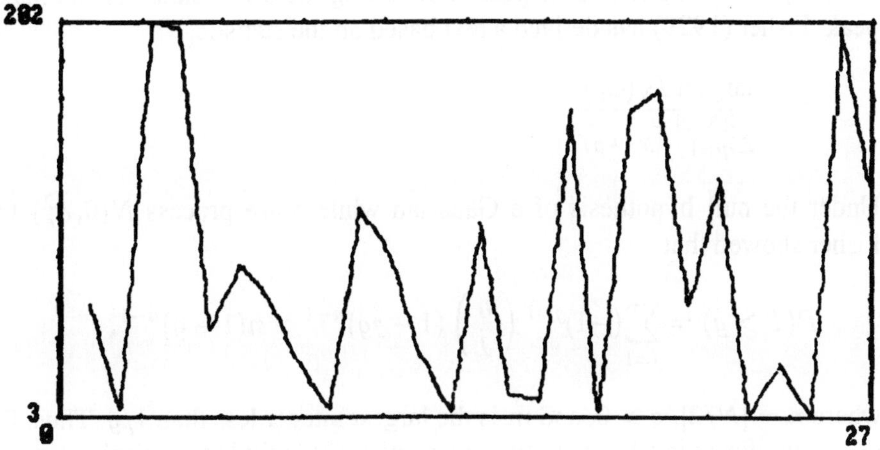

Figure 6. Normalized periodogram of inter-arrival times for circum-Pacific data.

Compute $I(\omega)$ at the frequencies

$$\omega_p = \frac{2\pi p}{N}, \quad p = 0, 1, 2, \ldots, \left[\frac{n}{2}\right].$$

Suppose that the data is generated from the model

$$X_t = \sum_{i=1}^{k} A_i \cos(\omega_i t + \phi_i) + e_t, \tag{1}$$

where phases $\{\phi_i\}$ are random having a uniform distribution, k is the number of periodic components, and $\{e_t\}$ are a sequence of uncorrelated random variables with mean zero and variance σ_e^2. Then

$$R_x(s) = \cos(X_t, X_{t+s}) = \frac{1}{2}\sum_i A_i^2 \cos\omega_i s + \sigma_e^2 \delta_{s,0},$$

where

$$\delta_{s,0} = \begin{cases} 1 & \text{if } s = 0, \\ 0 & \text{otherise.} \end{cases}$$

It can be shown that

$$E(I_N(\omega_p)) = \frac{N}{2} A_i^2 + 2\sigma_e^2, \quad \text{if } \omega_p = \omega_i,$$

$$E(I_N(\omega_p)) = 2\sigma_d^2, \quad \text{if } |\omega_p - \omega_i| > 0.$$

Thus, as ω_p approach each of ω_i, the ordinate of I_N (ω_p) is of order N and the periodogram exhibits a large peak. For testing the significance of the maximum peak, Fisher (1929) has derived a text based on the statistic

$$T = \frac{\text{largest } I_N(\omega_p)}{\sum_{p=1}^{[N/2]} I_N(\omega_p)}.$$

Under the null hypothesis of a Gaussian white noise process $N(0, \sigma_e^2)$ for X_t, Fisher showed that

$$P(T > g) = \sum_{j=1}^{m} (-1)^{j=1} \binom{n}{j} (1 - jg)^{n-1} \simeq n(1 - g)^{n-1}, \tag{2}$$

where $n = [N/2]$, $g > 0$, and m is the largest integer less than $1/g$. Thus, for any given significant level α, we can use Equation (2) to find the critical value g_α such that

$$P(T > g_\alpha) = \alpha. \tag{3}$$

Whittle (1952) suggested extended Fisher's test for the second largest ordinate based on the statistic

$$T = \frac{\text{second largest } I_N(\omega_p)}{(\sum_{p=1}^{[N/2]} I_N(\omega_p)) - \text{largest } I_N(\omega_p)}, \tag{4}$$

where the distribution in (2) is used but with n being replaced by $(n - 1)$. The procedure can be continued until a significant result is obtained. This leads to an estimate for k, that is the number of periodic components in the Equation (1). Note that, for this problem an alternative approach is the construction of a test based on standard regression analysis.

3. Periodogram Analysis of Intervals Between Arrival Times

Suppose that we have observed N occurrences at times $t_1, t_2, \ldots, t_N$. For example, these can be the arrival times of earthquakes. We can convert the data (i.e. arrival times) to an another time series by binding the data or measuring the intervals between arrival times, that is

$$X_j = t_{j+1} - t_j, \quad j = 1, 2, \ldots (N-1)$$

The advantage of this conversion is that the vast methodology available in time series literature can be applied using X_j's. If the arrival times are Poisson, then we know that $\{X_j\}$ is negative exponential and we can test for Poisson by testing for the exponential distribution. For a discussion concerning a Poisson process and its applications in earthquake risk evaluation see Daley and Vere-Jones (1988).

Now suppose that the arrival times are in the form of periodic (quasi-periodic) or regular cluster, then the intervals can be represented by a harmonic model (1) described in Section 2, and the method presented above can be used for detection and testing. This is described in the next section using a set of real data.

4. Application to Earthquake Data

In order to demonstrate the approach we consider arrival times of the earthquakes (with body wave magnitude greater or equal to 3) which have occurred in the central USA (lat. = 35°–39° N, long. = 87°–92° W) during the period 1920–1990. The data consists of 407 arrival times. See Dargahi-Noubary and Razzaghi (1994) for complete listing of this data. The inter-arrival times and their periodogram are presented in Figures 1 and 2 respectively. As can be seen, there are relatively large peaks at 155, 81, 106, 154, and 192.4 (days). However, none of these is significant at 5% significance level leading to rejection of a period component. Looking at the histogram of the interval times (Figure 3) one can comfortably conclude that an exponential distribution will fit the times. Hence, a Poisson arrival times cannot be rejected here.

Next we consider earthquakes with $M_S \geqslant 7.0$ in the circum-Pacific region as used in Kagan and Jackson (1991). For this data the time limits are from 1 January 1973 to June 1988. The latitude and longitude limits are respectively 67° N to 60° S and 30° W to 90° E. Taking 30 January 1973 as the baseline, we compute the number of days between two successive events. Figures 3, 4 and 5 present interarrival times, autocorrelation, partial correlation and periodogram for this data. Again, there are few peaks, and none are significant. The value of $P(T > G_\alpha) = \alpha$, that is Equation (3) is 0.946. This means that, once more a Poisson arrival times cannot be rejected.

5. Conclusion

A test based on periodogram analysis of time intervals between successive earthquakes is described for checking the cyclicity of earthquakes data. The results obtained for the two sets of data show that the existence of a periodic component can be rejected with a high confidence and that a Poisson arrival times cannot be rejected.

References

Bakun, W. H. and Lindh, A. G.: 1985, The Parkfield, California, earthquake prediction experiment, *Science* **229**, 619–624.

Daley, D. J. and Vere-Jones, D.: 1988, *An Introduction to the Theory of Point Processes*, Springer-Verlag, Heidelberg.

Dargahi-Noubary, G. R. and Razzaghi, M.: 1994, Earthquake hazard assessment based on bivariate exponential distributions, *Reliability Eng. System Safety* **44**, 153–166.

Epstein, B. and Lomnitz, C.: 1966, A model for the occurrence of large earthquakes, *Nature* **211**, 954–956.

Fisher, R. A.: 1929, Tests of significance in harmonic analysis, *Proc. Roy. Soc. Ser. A.* **125**, 54–59.

Kagan, Y. Y. and Jackson, D. D.: 1991, Seismic gap hypotheses: Ten years after, *J. Geophys. Res.* **96**, 21,419–21,431.

Kelleher, J. A., Sykes, L. R., and Oliver, J.: 1973, Possible criteria for predicting earthquake locations and their applications to major plate boundaries of the Pacific and Caribbean, *J. Geophys. Res.* **78**, 2547–2585.

Lomnitz, C.: 1985, Tectonic feedback and earthquake cycle, *Pure Appl Geophys.* **123**, 667–687.

Lomnitz, C.: 1994, *Earthquake Prediction*, Wiley, New York.

Nishenko, S. P.: 1991, Circum-Pacific seismic potential – 1989–1999, *Pure Appl. Geophys.* **135**, 169–259.

Schuster, A.: 1898, On the investigation of hidden periodicities with application to a supposed 26-day period of meteorological phenomena, *Terr. Mag. Atmos. Elect.* **3**, 13–41.

Sieh, K., Stuiver, M., and Brillinger, D.: 1989, A more precise chronology of earthquakes produced by the San Andreas fault in southern California, *J. Geophys. Res.* **94**, 603–623.

Sykes, L. R. and Nishenko, S. P.: 1984, Probabilities of occurrence of large plate rupturing earthquakes for the San Andreas, San Jacinto, and Imperial faults, California, 1983–2003, *J. Geophys. Res.* **89**, 59,095–59,270.

Whittle, P.: 1952, The statistical analysis of a seiche record, *J. Marine Res. (Sears Foundation)* **13**, 76–100.

Working Group on California Earthquake Probabilities, Probabilities of large earthquakes occurring in California on the San Andreas fault, 1988, U.S. Geol. Surv., Open File Rep., 88–398, 62 pp.

Natural Hazards **16:** 135–163, 1997.

Climate Change, Extreme Events and the Canadian Insurance Industry

RODNEY WHITE[1] and DAVID ETKIN[2]
[1]*Director, Institute for Environmental Studies, University of Toronto, 33 Willcocks Street, Toronto, Ontario, Canada M5S 3E8*
[2]*Environmental Adaptation Research Group, Environment Canada, University of Toronto, 33 Willcocks Street, Toronto, Ontario, Canada M5S 3E8 e-mail: david.etkin@ec.gc.ca*

(Received: 26 November 1996; in final form: 14 August 1997)

Abstract. At the same time that a scientific consensus has arisen that the world will most likely experience a changing climate in the near future, with more frequent extreme events of some weather hazards, the insurance industry, worldwide, has been hit with rapidly escalating costs from weather-related disasters. This conjunction of scientific belief and economic impact has raised the questions as to (1) whether more frequent extreme events have contributed to the rising insurance costs and (2) how will future climate change affect the industry?

Based upon historical data, it is difficult to support the hypothesis that the recent run of disasters both world-wide and in Canada are caused by climate change; more likely other factors such as increased wealth, urbanization, and population migration to vulnerable areas are of significance.

It seems likely, though, that in the future some extreme events such as convective storms (causing heavy downpours, hail and tornadoes), drought and heat waves will result in increased costs to the industry, should the climate change as anticipated.

Key words: climate change, extreme events, insurance.

1. Introduction

In a recent seminar at the University of Toronto, Angus Ross, Chief Agent of SOREMA Re Canada, observed that the five years from 1985 to 1989 produced 11 major insurance payouts from extreme atmospheric events, while the period from 1990 to 1995 produced 24 (Ross 1996). Not only were the events more frequent, they were also more severe in their impact, physically and economically. Along with earthquakes in populated areas, extreme weather events have the power to bankrupt insurance companies, and perhaps destabilise the industry globally. Hurricane Andrew, which, in 1992 inflicted insured losses of US $16.5 billion and economic losses of at least $25 billion, is credited with the bankruptcy of ten insurance companies.

Canada has not suffered from weather events as serious as Andrew, though hurricane Hortense struck Nova Scotia in September 1996, produced insurance claims of Cnd. $3 million, and was the first hurricane since Blanche (in 1975) to make landfall in Canada. The year also witnessed floods in Ottawa – Hull (insured losses of $20 million), hailstorms in Calgary and Winnipeg ($140 million and $120 million) and the Saguenay flood ($350 million). Furthermore, Canadian companies write policies outside the country, including for such vulnerable states as Florida and California, and, given the international nature of the industry the cost of policies written for Canada are affected by losses that companies endure elsewhere. Reinsurance – the coverage which the primary insurers buy for themselves – is an international business, and it is these rates which are most sensitive to billion dollar losses.

Though it is unclear whether or not the recent insurance losses are related to a statistically significant increase in extreme weather events, they do point out the sensitivity of the industry to weather. If, indeed, climate change will result in a greater number of weather disasters, then the insurance industry cannot afford to take a 'wait-and-see' position with regard to climate change. There are certain 'no regrets' options open to the industry, but these alone are not an adequate response to the potential dangers. A more pro-active approach is indicated, one that accepts probabilities, and does not wait for scientific certainty. The purpose of this paper is to assess what those probabilities are and what they might mean for the insurance industry, especially the property and casualty insurance companies. The paper has a Canadian focus; information is also drawn from the United States, and from Europe, especially the United Kingdom.

The purpose of this paper is to explore the implications of climate change for the insurance industry (with particular reference to Canada) by reviewing industry losses to extreme weather within the context of how these weather extremes may vary in the future.

2. Current Consensus on Atmospheric Change and its Implications

The Intergovernmental Panel on Climate Change (IPCC, 1995) recently co-ordinated an assessment of available knowledge, an exercise in scientific consensus building. Some of their key conclusions, though some controversy around these must be noted (Singer *et al.*, 1997), are

(i) there is high confidence that human activities have significantly increased the concentration of radiatively active gases (RAGS) within the atmosphere, and have therefore enhanced the natural greenhouse effect. This enhancement will continue to increase as human emissions of greenhouse gases continue at more than one-half of present global emissions. There is a real and significant risk that global scale climate changes, dangerous to ecosystems and human society, will result from these increases;

(ii) the balance of evidence from analysis of past climates suggest that there already exists a discernable human influence on the global climate system;

(iii) prediction with respect to the magnitude, rate, and regional characteristics of future climate change remain largely uncertain, although broad conclusions with respect to continental and latitudinal response can be made with some confidence;

(iv) in general, predictions on the changes in probabilities and intensity of most extreme events as a result of an enhanced greenhouse effect remain uncertain, although there are indications that more rainfall will occur as extreme events, and that heat waves and droughts may likewise become more frequent and more severe.

Usually, in science, as one becomes more reductionistic ones confidence improves. This is because the number of variables is reduced. In the case of climate change, the greatest confidence exists for globally averaged values, which can be correct for a number of climate configurations; regional climate change is very dependent upon the specific configuration of the atmospheric-oceanic system.

Though uncertainty is large, the "least likely scenario is considered to be no climate change" (Arrhenius and Waltz, 1990). Ignoring climate change as a factor in a risk assessment means accepting the following assumption: 'Large changes in radiatively active gases (CO_2, methane, etc.) which are fundamental to the energy balance in the atmosphere, this energy balance being a principal boundary condition on climate, *will not* result in a change in climate'. Scientific evidence runs contrary to this assumption. On this basis, a 'wait-and-see' attitude is not rationally supportable. The question then becomes, how do we estimate and manage the risks associated with climate warming? Unfortunately, there is no really satisfying answer to this question.

3. The Relationship Between a Warmer World and Extreme Weather Events

3.1. THE NATURAL HAZARDS GAME

There is a difference between natural hazards and natural disasters. A tornado is an example of a natural hazard; a tornado disaster occurs when a severe tornado destroys part of a town, causing significant loss of life and property, often beyond the ability of the local community to recover from, without assistance. Society usually copes well with 'run of the mill' hazards. Disasters (such as the Kobe earthquake in Japan or Hurricane Andrew in Florida), though rare, can be devastating.

In one sense, coping with natural hazards is like a game (the Natural Hazards Game). Once you know the rules, you can devise a winning strategy. If the chances of good or bad events occurring are fairly well known, then the tradeoffs can be calculated with some level of comfort.

But what if after playing the Natural Hazards Game for some time, it turned out that the rules were about to change? Not only that, exactly when and how they were going to change, or precisely what the new rules would be was not known? Even worse, the new set of rules would not be constant, but would evolve over time, and not planning for them could result in 'losing' the game. That is what climate change does to the Natural Hazards Game.

3.2. CLIMATE CHANGE AS A HAZARD

3.2.1. *How Climate Change is Hazardous*

Climate can be represented as a probability function, or a histogram of the frequency with which weather events occur. Depending upon the weather element selected and the location, a variety of probability distributions are possible. For example, temperature is often represented by a normal distribution (Wigley, 1988), storm waves by a Rayleigh (Khandekar and Swail, 1995) and precipitation by a lognormal. These distributions are characterized by their means and variances.

Though many extreme events occur at the tails of the probability function of a single homogeneous population, some events are 'hybrids', and result from the unfortunate convergence of more or less unrelated events. One example is Hurricane Hazel, a tropical storm which was picked up by a strong mid-latitude jet stream and gained new energy. Another is the Alberta flood of June, 1995, where an intense rainfall event coincided with the rapid melt of a deep winter snowpack.

There are several ways in which climate can change:

(i) The mean of a distribution could shift towards more positive or negative values. An example of this is the global warming of around 0.5 °C over the past century.

(ii) The variance of the distribution could change. Examples of this are decreased variability of the daily temperature range over the U.S., Soviet Union and China (Karl and Knight, 1995), or the increased variability of rainfall over the Sahel (Hulme, 1992).

(iii) A distribution could become more or less skewed. An example of this is presented in Lambert (1995) who found a decrease in weaker storms but an increase in intense storms in a doubled CO_2 GCM experiment, where the variance decreased by about 20% and the skewness by about 25%.

(iv) The incidence of hybrid events could change. For example, if the mid-latitude jet stream migrates northward in a warmer climate but there is no change in the areas of hurricane occurrence, then hybrid events such as Hurricane Hazel might become less likely. The likelihood of a hybrid hazard depends not only on the probability distribution of the relevant components, but also on their geographical distribution.

Rates of climate change can be linear or non-linear. A 'sudden' change can occur due to the superposition of more than one forcing (constructive interference of waves), or because the climate system reaches a bifurcation point, and switches into a different stable mode. An example of the latter would be a switch in the ocean circulation patterns from the current conveyer belt to one with greater symmetry between the Atlantic and Pacific circulations (Manabe and Stouffer, 1988). There is mounting empirical evidence that climate tends to exist in several states or modes, and switches rapidly between them (Broecker, 1995).

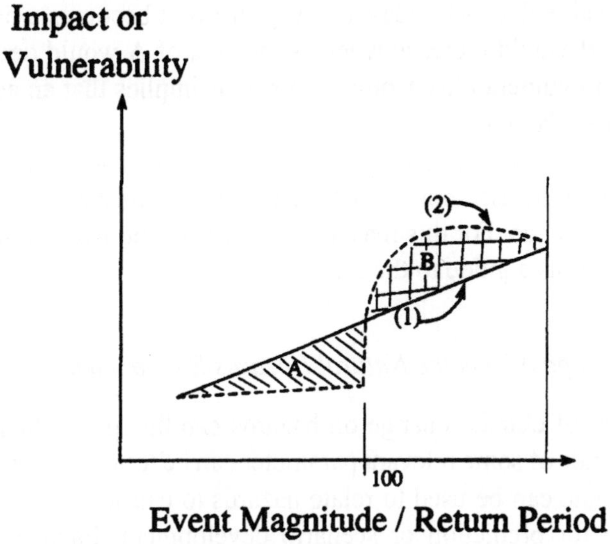

Figure 1. Illustration of the effect of mitigation on vulnerability.

Though a change in mean, such as average precipitation, can be hazardous in the sense that it represents a change from what society has adapted itself to, it is possible that the greatest impact of climate change will be due to increases or decreases in the frequency of extreme events. A discussion of the effect of a changing climate on extreme events can be read in 'Climate Models-Projections of Future Climate' (IPCC, 1995), the conclusions of which are in agreement with this discussion.

The relevance of climate change to natural hazards is that the frequency of extreme events may increase, and thus current adaptation measures may prove to be less adequate than at present. This is illustrated in Figure 1.

More extreme events have larger return periods (ie., they occur more rarely), which is illustrated by increasing values along the X axis. The impact of these events increases with return period, that is society is more vulnerable to rarer events of greater magnitude. Line (1) depicts a situation where society has not mitigated the hazard, and is exposed to it at all levels of magnitude. In this situation the vulnerability of society and the resulting impact of the event increases with the magnitude of the event. Where mitigation has occurred (for example, dams or levees built to protect society against the 100 year flood, illustrated by the dashed line #2), vulnerability remains relatively low up to that return period. However, if increased development has occurred in, for example, flood prone areas, then vulnerability may well have increased for events with return periods longer than 100 years, hence the large increase after the 100 year event. The areas of 'A' and 'B' in the figure represent the changes in vulnerability between the natural system and the protected system. Hopefully the area of A is greater than that of B, or the overall vulnerability of society has increased. If, due to climate change, the return period of some design value decreases (say, a 100 year event becomes a 50 year event), then the area of B would increase whereas the area of A would decrease, and an overall increase in vulnerability would occur. This implies that an adapted society could become maladapted.

Any assessment of risk, such as done by the insurance industry for natural hazards, that is based upon the assumption that the climate in constant is likely to be in error, perhaps seriously so. For this reason the insurance industry should not simply use past experience as a guide to the future.

3.2.2. *How Climate Change Affects the Natural Hazards Environment*

Understanding the effect of climate change on hazards can be approached from several perspectives. Trends of some relevant parameter can be tested for statistical significance, empirical data can be used to relate hazards to causal phenomenon, or models can be used for prediction or scenario development. Each of these approaches have value, and can be used to make inferences about how the future may or may not evolve.

Tropical Cyclones. It has been suggested that tropical cyclones and hurricanes might become more frequent or intense in a warmer climate. Tropical cyclone frequency can be related to six physical parameters (Ryan *et al.*, 1992), three dynamic (low-level relative vorticity, the Coriolis parameter and the vertical wind shear), and three thermodynamic (the ocean thermal energy, moist instability and relative humidity). The basis for arguments of increasing frequency rest upon prognoses of increasing sea surface temperatures (SST) or ocean thermal energy.

Since the mid-1970s there has been a decrease in the number of intense Atlantic hurricanes (Landsea *et al.*, 1996), though globally there seems to be no trend. The Atlantic decrease has been correlated with more frequent El Nino episodes and Sahel rainfall. There is some evidence, however, that more frequent hurricanes in 1995 and 1996 represent a return to a cycle of increased frequency of activity, after the relatively quiescent decades of the 1970s and 80s. Pielke and Landsea (1997), noting that hurricanes are the costliest natural disasters for the U.S., found no trend in damages when data was corrected for inflation, population and changes in wealth.

Emmanuel (1987) looked at this issue by modelling the tropical cyclone as a Carnot heat engine, and concluded, using thermodynamic arguments that emphasize SST, that warmer SST in a doubled CO_2 world would increase the maximum sustainable pressure drop in the storm, potentially increasing storm intensity by 40–50%. He also suggested the possibility of ultra-powerful hurricanes (hypercanes) if the SST rose by 6 °C or more.

Idso *et al.* (1990) used empirical data to support the hypothesis that hurricanes would not be more frequent in a global climate warmer by 1/2–1 °C, and might be less intense.

Broccoli and Manabe (1990) examined tropical cyclone frequency in an equilibrium CO_2-doubled GCM, and found a large increase in storms when cloudiness was constrained to climatology, but a lower frequency of storms when cloud feedback is allowed. Interestingly, SST increases were larger in the cloud feedback case, which points to the importance of factors other than SST as important to cyclone development, especially cloud feedbacks. In another GCM experiment, Haarsma *et al.* (1993) found a 50% increase in the number of cyclones, with relatively more intense ones.

Evans (1993) examined empirical data relating SST to cyclone intensity, and found no obvious relationship, concluding that the availability of increased ocean thermal energy is not the limiting factor for the intensity of tropical cyclones, though it might define an upper bound.

Ryan *et al.* (1992) used the Gray Seasonal Genesis Parameter (SGP) to examine cyclone frequency changes from a GCM, and found large increases resulting from the increased SSTs, though they note that it is the dynamical factors that determine cyclone frequency, once the minimum thermodynamic factors are satisfied. This study points to the danger of using empirically derived relationships and extrapolating them to different environments.

As part of the WMO/ICSU program on tropical cyclone disasters, a workshop was held 22 November–1 December 1993. The participants (Lighthill *et al.*, 1994) concluded through an examination of observational data that 'though the possibility of some minor indirect effects of global warming on tropical cyclone frequency and intensity cannot be excluded, they must effectively be 'swamped' by large natural variability', and felt that the use of climate models to assess changes in cyclone frequency was not at a useful stage. In a critique of Lighthill *et al.*, 1994, Emmanuel (1995) argues that though the frequency might be unchanged, the arguments presented regarding intensity are flawed, and that the potential for more severe storms exists as a result of a climate warming scenario.

Extra-Tropical Storms. It has commonly been argued that since polar latitudes are expected to warm more than mid or tropical latitudes, then the decreased temperature gradient will result in weaker mid-latitude storms, where the energy source is baroclinicity in the atmosphere. This assumption is complicated by the prediction of an increased temperature gradient in the upper troposphere, as a result of a large warming in the tropical upper troposphere. It is not clear therefore, whether baroclinic storms will become stronger or weaker (Held, 1993), though some theoretical calculations suggest that the lower troposphere gradients dominate. The effect of moisture also complicates the issue. Latent heat release should increase in a warmer climate, thus strengthening storms, though the transport of latent heat in large scale eddies from the tropics to the pole may reduce the requirement for balancing the global energy budget, the ultimate driving force behind mid-latitude storms. Changes in the circulation pattern may well alter storm tracks, an effect which on a regional scale may swamp other considerations Balling and Lawson

(1982) found a shift in winter circulation patterns over North America in the early 1950's, from predominantly zonal to meridional, a change that would have major impacts on storm tracks. They also noted that the interior plains and northeast quarter of the U.S. appear to be most sensitive to the change in circulation. Prediction of storm tracks in a warmer climate remains a major challenge (Held, 1993).

Agee (1991) examined storm frequencies during periods of hemispheric warming and cooling, and found statistically significant linear relationships between the two. During periods of warming, cyclone frequency increases, while during periods of cooling, it decreases. In the 1950 to 1975 cooling period, storm frequencies dropped by 30%, while in the 1905 to 1940 warming period, they increased by around 19%. Stein and Hense (1994) found a higher frequency of extreme storms in the North Atlantic winters since 1988/89 than at any time since 1880. Lambert (1996) found an increase in severe winter cyclones near the Aleutian Low and the Icelandic Low after 1980.

Lambert (1995) using the CCC GCM, found a 4% decrease in cyclones in the northern hemisphere though the frequency of intense cyclones increased. Lambert hypothesises that the latent heat effect is responsible for the greater number of intense storms. No change in storm tracks was evident. A few areas showed increased frequencies, such as off Cape Hatteras, over Hudson Bay and west of Alaska. These results are similar to Rowntree (1993) who found a 40% increase in Atlantic gales, though fewer intense storms over eastern North America. Hall *et al.* (1994) and Carnell *et al.* (1996) found an intensification and northward shift of storm tracks.

Evidence on how storminess will change in a warmer climate is conflicting, and conclusions, especially regarding severe storms must be viewed as uncertain.

Convective Storms. Convective storms (severe thunderstorms producing hail, lightning, tornadoes, heavy rain and strong winds) remain a particularly difficult issue for GCMs because of their small scale. Intuitively, one would expect more frequent and more intense convection since a warming surface and a cooling stratosphere in mid-latitudes will destabilize the troposphere. Mitchell and Ingram (1990) found deeper convection in a GCM forced by doubled CO_2. Noda and Tokioka (1989) in another GCM experiment found that global precipitation increased in a similar experiment, though the precipitation area decreased. Both these studies suggest more frequent intense convective rainfall in a warmer climate. Thus the frequency of both floods and droughts may increase (Meehl, 1993). The issue of dynamics also needs to be addressed though, and it is not clear how changing wind fields will affect severe storm frequency.

Price and Rind (1993) found, using a GCM, that in a $2\times CO_2$ climate with a 4.2 °C warming global cloud-to-ground lightning strokes will increase by 72% over continental regions.

Etkin (1995) in an empirical examination of tornado occurrence in the prairies of western Canada found that tornado frequency is greater in warmer springs and summers. This implies that the number of tornadoes may increase there as a result of climate change. Interestingly the number of observed tornado events in southern Ontario was low in some hot summers. Mike Leduc (personnel communication) thought this might be because the polar front migrated to the unpopulated regions of northern Ontario where events go largely unreported.

Griffiths *et al.* (1993) discuss the difficulties in assessing convective changes and suggest three approaches, one by correlating data with general circulation indices, a second by using models to compare the frequency of pre-conditions to severe storms, and a third by comparing present and projected climatologies of synoptic situations.

Hogg (1996) examined hydrological extremes in Canada, and found a slightly positive but not significant trend. This may not be too surprising, in light of Ross and Elliot (1996) who found significant increases in precipitable water in the U.S. but not in Canada. Also, Karl *et al.* (1995) found that the proportion of precipitation contributed by extreme one-day summer events in the U.S. increased significantly from 1910 to 1990.

It seems likely that the number of convective events will increase in most areas in a warmer climate, due to the thermodynamic changes. This assumes that the dynamical contribution to convective storms (associated with wind fields) does not diminish and thereby compensate for temperature and humidity changes.

Weather Disasters and Temperature. Changnon and Changnon (1992) did a study of trends in weather disasters in the U.S., attempting to correlate them with temperature trends. A disaster was defined as a storm which produced more than $100 million in losses (1991 dollars), which provided 142 cases from 1950 to 1989. These storms produced $66 billion in losses. Included were hurricanes, winter storms, convective storms and windstorms. The 10 worst storms, 7% of the disasters, accounted for almost half of the total losses. These figures become more extreme if hurricane Andrew is included. Figure 2 shows five year averaged trends of storm frequencies against annual temperature and precipitation, which show relatively more storms causing damage after 1970, mainly due to thunderstorm-related events. These events occurred mainly in the southern and eastern U.S. Figure 2 is particularly interesting, as it shows a correlation between weather disasters and the U.S. annual temperature. The correlation was strongest with convective storms, followed by winter storms. Hurricane relationships were weak. The authors conclude that warmer climates may cause higher extreme storm frequencies in the eastern two-thirds of the U.S.

Extreme Temperature Events. In a warmer climate, heat waves would become more frequent, while cold waves would become less so. Evidence suggests that even a warming of a couple of degrees Celsius can have a major impact on this hazard, due to non-linearities in the system.

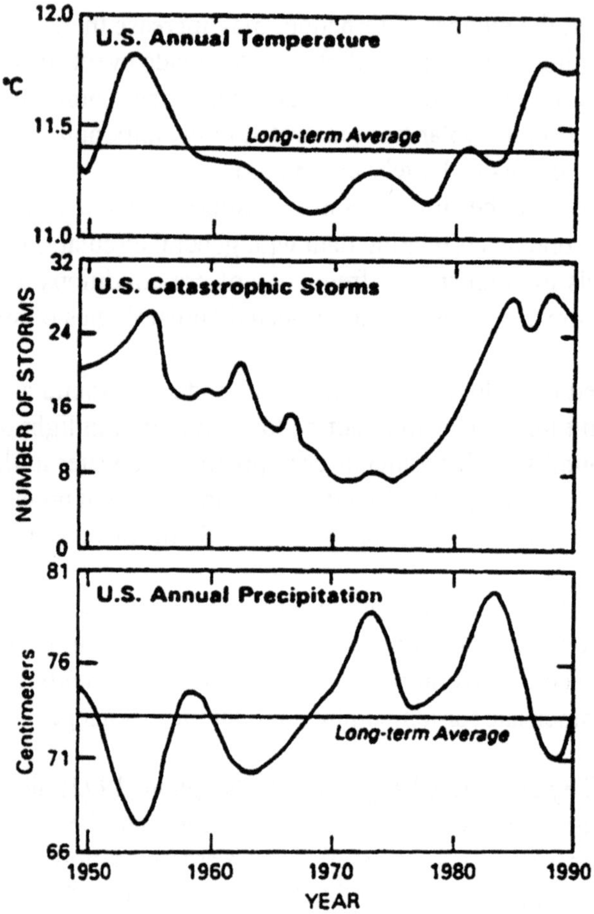

Figure 2. Five-year moving average of national U.S. frequency of weather disasters and annual mean temperature and precipitation (Changnon and Changnon, 1992).

The frequency with which extreme temperature events occur has been analysed by Mearns *et al.* (1984), Wigley (1988) and Katz and Brown (1992). Mearns *et al.* (1984) note the strong non-linear relationship between changes in the mean and changes in the probability of extremes, which may be the principal way in which climate change is felt. They found large changes in the likelihood of heat waves at Des Moines (by a factor of 3), with relatively small changes in mean temperature (1.7 °C). As well, changes in the variance and/or autocorrelation of a temperature time series can significantly affect the probability of extreme events, potentially being as or more significant than changes in the mean.

Wigley (1988) found that risk is extremely sensitive to changes in the mean (assuming that the extreme events come from the same parent population). He found, for example, that for an event with a 10% risk of occurring in 100 years, that risk is increased to 90% if the mean were increased by 0.02 standard deviations per year during the 100 year timeframe (assuming a normal distribution). Thus, if

GCM models are correct, then rare events will become commonplace before the middle of the next century.

Katz and Brown (1992) analysed the sensitivity of extreme events to changes in the mean and standard deviation (for a normal distribution), and found that extreme events are more sensitive to variability than to average, and that this sensitivity becomes greater the more extreme the event. This conclusion was also noted by Barrow and Hulme, (1996). In an analysis of temperature extremes, a 0.5 °C change in mean resulted in a 35% increase in probability of daily exceedence of 38 °C, while the same change in standard deviation results in a 71% increase. This occurs because the sensitivity to mean increases linearly while the sensitivity to standard deviation increases quadratically. They conclude that any climate change impact study that holds variability constant while only changing the mean may not be realistic. The question of how to incorporate variability issues in climate change studies remains problematic; for example, Skaggs *et al.* (1995) demonstrate how temperature variance changes over time for the eastern U.S.A. Spatial and temporal analogues may provide insight regarding this issue. It must be noted though, that changes in the mean are expected to be more significant than changes in variance, according to climate models (IPCC, 1995 – Ch. 6.)

Zwiers (1994) found significant changes in the 10 year return period maximum temperature at 2 m, ranging from 4–6 °C over all continents except the Antarctic.

Hennessy and Pittock (1995) using a global warming scenario of 0.5 °C found 25% more days over 35 °C in summer and spring at Victoria, Australia, and 50-100% more in a 1.5 °C warming scenario.

Mearns *et al.* (1995) examined temperature variability and diurnal range in a nested GCM, and compared their results to others. They found, on average, a decrease in daily temperature range though with large regional and seasonal variations, mainly resulting from changes in absorbed shortwave radiation. The largest change in the variability of daily mean temperature was a decrease in winter in high latitudes. Changes in variability are mainly explained by changes in atmospheric circulation patterns (e.g., storm tracks) though changes in the Arctic ice regime are important for high latitudes. These results broadly agree with several other authors, though they note that there are considerable differences between models, and that research is needed for diagnosis.

Evidence points to much more frequent heat waves as a result of climate warming. Similarly, cold waves would become rarer.

Floods. Canada is subject to several different kinds of floods, including rainstorm floods, ice-jam floods and snow-melt floods. Rainstorm floods result from synoptic, meso- and microscale events, and can be of particular concern to urbanized areas, where the impermeable surface increases runoff. A shorter winter season may result in a reduced snow-pack in many areas and thereby a reduced risk of snow-melt and ice-jam floods, though greater snowfalls might compensate. Leathers and Ellis (1996) found large increases in lake-effect snowpacks to the lee of the Great Lakes.

Floods can also occur along coastlines, from storm surges and tsunamis. Rising sea levels associated with climate warming is likely to exacerbate the impact of storm surges, though there are few parts of Canada, aside from Richmond, B.C., with large vulnerable areas near sea-level.

Slivizky and Morin (1996) examined spring runoff in Quebec using four GCMs and found increases ranging from 1.6–13.3% as a result of increased snowmelt. These results suggest possible future increases in flooding during the spring.

Mainly, concern about increased flooding in a 2 × CO_2 world result from the fact that warmer atmospheres can hold more moisture, and precipitation is expected to increase as a result. As well, the precipitation is expected to become more convective in nature, and therefore more intense over smaller areas, which suggests greater flooding problems. In Canada, concern regarding floods was highlighted in 1996 and 1997, when severe flooding in Saguenay, Quebec (due to heavy rainfall) and southern Manitoba (due to the rapid melt of a heavy snowpack) occurred. The Saguenay flood may prove to be the costliest natural disaster thus far to the Canadian insurance industry, while the Manitoba flood has large financial implications for the public sector.

Gordon *et al.* (1992) while noting that GCMs cannot provide meaningful quantitative estimates of how extreme rainfall events may change, note that their GCM shows a marked increase in convective rainfall events and a mid-latitude decline of non-convective events. The frequency of large rainfall increased (with return periods decreasing by around a factor of two for the central U.S. but by up to five elsewhere) while the frequency of light rainfall days decreased for all regions, especially in mid-latitudes. These results are similar to Noda and Tokioka (1989), Hansen *et al.* (1988), and others. Also, Whetton *et al.* (1993) found increased extreme rainfall events from a GCM run for Australia, though noting the significant caveats with which such results must be treated.

This increase in variability (resulting from more favoured convection) suggests potentially large changes in the probability of extreme events, as discussed by Katz and Brown (1992). In one example by Smith (1993), a 25% increase in half hour rainfall intensities for Sydney changed the 1 in 100 year event into a 1 in 17 year event. While noting the severe limitations of estimating changes to flood probabilities, he notes that for Australia, there is a consensus that the frequency of extreme floods will increase. He also shows that changes in the occurrence of low probability extreme events result in relatively large increases in damage. In fact, historically in Australia, damage caused by greater than 1 in 100 year events is about equal to all the events below that return period.

Rind *et al.* (1989), Wilson and Mitchell (1987) and Parey (1994), contrary to other results, did not find evidence of increased heavy rainfall in a GCM experiment.

Lawford *et al.* (1995) found no clear evidence of historical trends that indicate changes to extreme flood events in Canada, though the data is suggestive that Alberta may be experiencing more heavy rainfall storms since the 1960s.

In summary, climate models suggest an increase in flooding events, as a result of a trend towards more convective precipitation and greater atmospheric absolute humidity.

Drought. If floods are the Yin of the precipitation hazard, then drought is the Yang. The concern is that if precipitation becomes more convective with an increase in heavier events then the number of dry days will increase and drought will become more severe (IPCC, 1995). This could be exacerbated by increases in potential evapotranspiration due to higher temperatures.

An interesting paper by Hughes and Brown (1992) indicates that central California has had fewer droughts in the period from 1850 to 1950 than at any time in the last 2000 years. This suggests that the current climate is anomalously benign, and that increased drought frequency in the future is not unlikely for that region. Vance (1991) found that drought on the northern Great Plains are not cyclical, but rather that intervals of intense drought are interspersed between longer periods when drought is rare.

Oladipo (1993) analysed drought in northern Nigeria, and found a statistically significant abrupt transition towards lower precipitation in the Sahel region beginning in 1968.

Whetton *et al.* (1993) notes that GCMs cannot adequately simulate soil water balances and examined drought in Australia using a stand alone soil water balance model. His results found unclear signals about the soil water regime, and he was not able to draw any conclusions about the future probability of drought.

Some GCM studies show reduced summer soil moisture values over the mid North American continent, suggesting more frequent droughts, though Maybank *et al.* (1995) indicate that the trend is not clear.

Cubash *et al.* (1995) found a doubling from 1 to 2% in the frequency of 3 month droughts in central North America in a $2 \times CO_2$ run. Wetherald and Manabe (1995) attribute increased drought in the models to a reduction in mid-latitude soil moisture as a result of more evaporation.

Though the evidence is still not clear, there is a reasonable basis for concern that the frequency of meteorological droughts will increase in the future.

Other Hazards. Various other hazards are tied to the more primary ones discussed above. For example, wildfires are a function of temperature, the precipitation regime and lightning, and are quite sensitive to climate variability and change. This can be seen by examining fire damage in Canada in the 1980's and 1990's, which increased significantly above 1920–1980 levels (Kurz and Apps, 1996). Street (1989) depicts a longer and more severe forest fire season in Ontario with climate warming, with a shift in timing towards later in the season for the most severe period, given current GCM outputs. Storm surges and storm waves result from ocean or lake based storms. It is not clear how these hazards will evolve if the climate warms (Khandekar and Swail, 1995) due to uncertainty about storm intensities. Karl *et al.*

Table I. Trends in hazards with climate warming

Hazard	Trend with Climate Warming	Confidence
Tropical Cyclones	No significant trend foreseen	Trend somewhat controversial
Extra-Tropical Storms	Conflicting evidence	
Convective Storms	More frequent and severe	High
Heat Waves	More frequent and severe	Very high
Cold Waves	Less frequent	Very high
Floods	More frequent	High
Drought	More frequent, possibly prolonged	Moderate

(1995) developed a climate extremes index (CEI) for the U.S. and found that the trend supports the argument that the climate has become more extreme since about 1976 (mainly as a result of changes in precipitation, not temperature), though the pattern does not lie outside historical variability.

3.2.3. *Summary*

Table I summarizes the scientific consensus of the above discussion. Trends are most likely to occur in the frequency of heat and cold waves, followed by more frequent convective storms and floods. Drought is also expected to become more frequent, though confidence is not as high for this hazard.

3.3. THE COSTS OF NATURAL HAZARDS

In order to get an appreciation for how increasing frequencies of natural hazards might affect Canadians, one must look at their historical and current costs. Many of these costs, both social and economic, are not well documented, and are often unobtainable. As a result, only a partial inventory can be reasonably attempted. It is not argued that any upward trends in Canada necessarily result from an increased frequency of hazardous events – changes in population, wealth, government policy etc can probably account for most or all of the trends shown. As noted above, in recent years the number of insured losses over $4 million have certainly increased. From 1980 to 1985 there were 11 major losses, while from 1990 to 1995 there were 24 (Ross, 1996). World-wide, the relationship between increased costs and events may exist; Paul Kovacs (personal communication) indicates that analyses carried out by the insurance industry conclude that the observed increase in insured costs, in part, can only be explained by more frequent and more severe storms.

Six basic strategies for adapting to climate have been identified (IPCC, 1995), these being: (1) prevention of loss, (2) tolerating the loss, (3) spreading or sharing the loss, (4) changing use or activity, (5) changing location and (6) reconstruction. North American society seems to have put particular emphasis on loss prevention through technology (e.g., dams), sharing the loss (e.g., insurance) and restoration.

This emphasis is based upon the debatable assumption that current loss rates are sustainable, and that social and other costs resulting from extreme events beyond design safety factors are not prohibitive. The costs of natural hazards are a combination of the cost of adapting to prevent damage plus the costs associated with damage when adaptations fail or extreme events overwhelm our protective systems. Recent experience has suggested that the above assumptions may not be valid, as the social costs of natural hazards world-wide has escalated manyfold in the past 15 years.

In the United States, it is estimated that since 1992 natural hazards cost about $1 billion per week (National Science and Technology Council, 1996). In the period from 1980 to 1997, there have been 28 U.S. weather related disasters that cost over $1 billion (Table II) that have resulted in many thousands of deaths and huge financial losses. The greatest costs in terms of dollars and deaths have been due to heat waves and droughts (a hazard which seems more likely to be increased by climate change than any other), though some hurricanes and floods have been almost as severe. Of the total costs, 53% are from droughts and heat waves and 23% are from flooding. Thus 76% of the costs are from hazards that are considered most likely to increase with climate change.

Insurance Bureau of Canada. Summaries of costs to the insurance industry are provided by the Insurance Bureau of Canada. Figure 3 shows the costs of multiple major payouts, from 1983-1996 (Source: Insurance Bureau of Canada). Data from 1995 and 1996 are preliminary. This data does not include the costs of events less than about $4 million, and therefore the true costs are much greater than those shown in this figure by an unknown amount. Hail (Figure 4) has caused the most damage (about $450 million – 41% of the total), followed by tornadoes (23%), flood (21%), storm (8%) and wind (6%). Though not included on the graph because data collection is still in progress, there appears to have been 9 events in 1995 (Alan Pang, personal communication) which include significant damage from flood, hail, thunderstorms, wind and Hurricane Hortense. Two hailstorms in Alberta and a thunderstorm in Ontario each are estimated to cost over $25 million. Hortense is expected to cost about $3 million. In July, 1996 hailstorms in Calgary and Winnipeg are estimated to cost around $295 million in total (Alan Pang, personal communication). The recent Saguenay flood is currently estimated at $350 million and will increase the importance of flood relative to other costs.

Crop Insurance. (Sources: British Columbia – Ministry of Agriculture, Fisheries and Food, Alberta Hail and Crop Insurance Corporation, Ontario Ministry of Agri-food and Rural Affairs, Manitoba Crop Insurance Corporation, New Brunswick Crop Insurance).

Provinces incur costs due to crop damage from hail, flood, drought and a variety of other hazards. Figure 5 shows the average annual provincial costs in unadjusted dollars. Saskatchewan incurs the greatest costs, almost $100 million per year. Crop

Table II. Weather related natural disasters in the U.S. where the cost exceeded $1 billion US (1980–1997) – ordered by rank in economic cost

Event	Rank by cost	Date	Economic cost ($ billions)	Lives lost
Drought/Heat Wave	1	Summer 1988	40	5000 to 10 000
Hurricane Andrew	2	August 1992	25	58
Drought/Heat Wave	3	June–Sept. 1980	20	1300
Midwest Flooding	4	Summer 1993	15 to 20	48
Hurricane Hugo	5	Sept. 1989	7.1	57
Hurricane Fran	6	Sept. 1996	>5	37
Drought – southern plains	7	Fall 1995	>4	0
Texas, Louisiana, Mississippi Flooding	8	May 1995	>3	27
California Flooding	9	Jan–March 1995	>3	27
Southeast Ice Storm	10	Feb. 1994	>3	9
Storm/Blizzard	11	March 1993	>3	270
Blizzard/Flooding	12	January 1996	3	187
Flooding – west coast	13	Winter 1996/97	2–3	36
Hurricane Opal	14	October 1995	2–3	21
Florida Freeze	15	Dec. 1983	2	0
Hurricane Allicia	16	Aug. 1983	2	21
Hurricane Iniki	17	September 1992	1.8	6
Hurricane Bob	18	August 1991	1.5	18
Hurricane Juan	19	Oct.–Nov. 1985	1.5	63
Flooding – northern plains	20	April–May 1997	1 to 2	11
Nor'easter 1992	21	December 1992	1 to 2	19
Hurricane Elena	22	Aug.–Sept. 1985	1.3	4
California Wildfires	23	Fall 1993	>1	4
Flooding/Tornadoes Mississippi and Ohio valleys	24	Dec. 96/Jan. 97	1	67
Flooding – pacific northwest	25	February 1996	1	9
Texas Flooding	26	October 1994	1	19
Tropical Storm Alberta	27	July 1994	1	32
Drought/Heat Wave	28	Summer 1993	1	unknown

Source: National Climate Data Center, 1995.

losses for Manitoba runs around $24 million per year, and are detailed in Figure 6, which shows the costs by hazard. Drought is the major hazard exceeding $250 million from 1966-1994 (38% of the total), followed by excess moisture (21%), hail (16%), heat (12%), frost (9%) and wind (2%). Other hazards are much smaller.

Ontario Road Accidents. (Source: Ontario Ministry of Transportation)

Data for Ontario road accidents was only available for 1991 and 1992. Figure 7 shows the weather related 1992 statistics, that account for about one-third of all

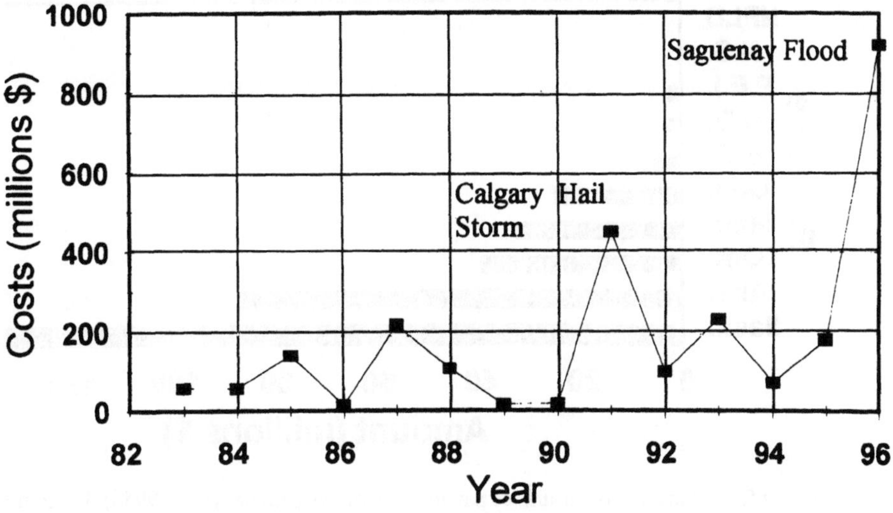

Figure 3. Weather related insurance costs (Major Multiple Payouts, 1995 $s).

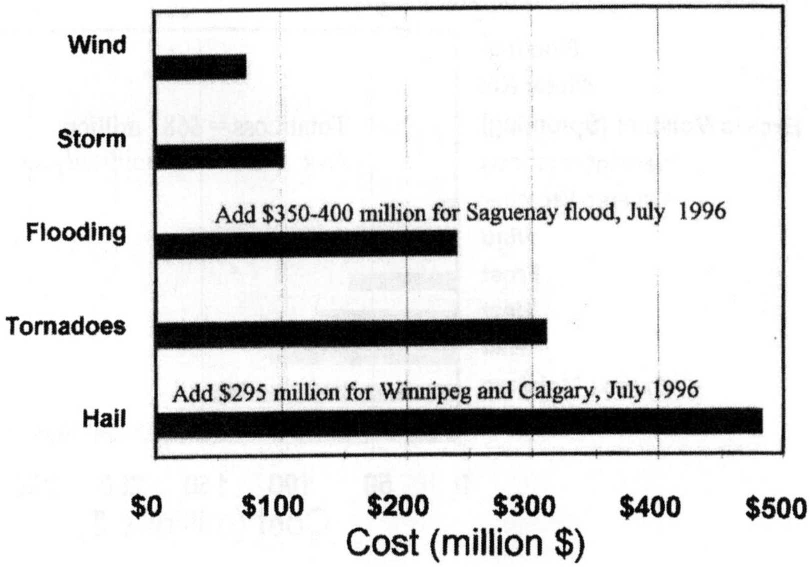

Figure 4. Weather related insurance costs (1984–1994) from major multiple payouts (1995 $s). Source: Insurance Bureau of Canada.

accidents. There were 298 fatalities, over 23 000 personal injuries and over 72 000 property damage cases. This does not include minor incidents not reported to the police. The majority of the weather related accidents resulted from wet conditions (62%), followed by snow (16%), ice (15%), slush (6%) and mud (0%). It is clear that weather-related car accidents are of great significance.

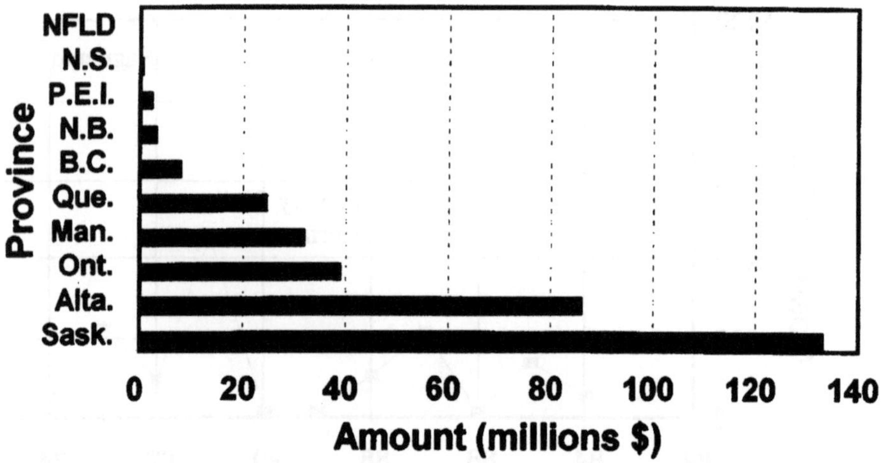

Figure 5. Provincial crop insurance payments. Average paid per year (1995$). Programs began as early as 1959 in Manitoba and as late as 1974 in NB. The average annual payment is $328 million. Sources: British Columbia Ministry of Agriculture, Fisheries and Food, Alberta Hail and Crop Insurance Corporation, Ontario Ministry of Agri-food and Rural Affairs, Manitoba Crop Insurance Corporation, New Brunswick Insurance.

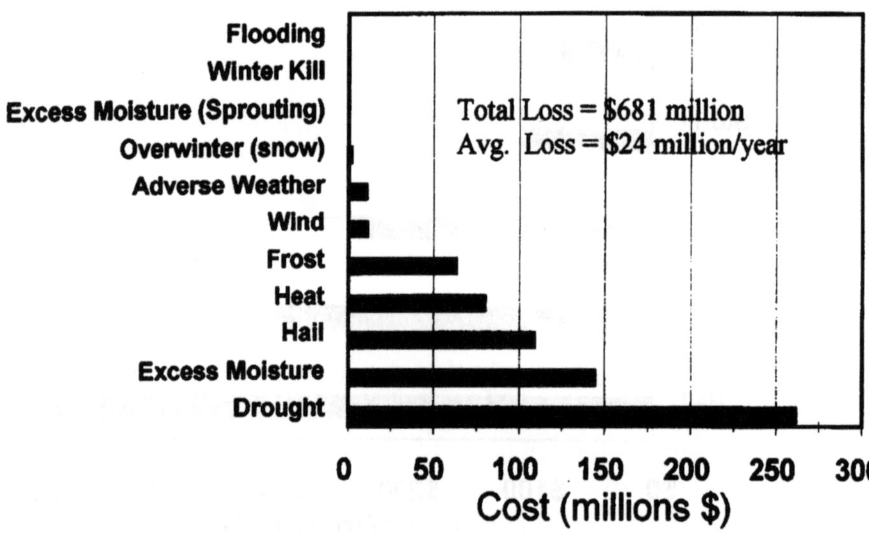

Figure 6. Manitoba – crop loss by cause 1966–1994 (dollars not corrected)

Food for Thought. It cannot be demonstrated that increased costs in the Canadian insurance industry result from more frequent or more severe natural hazards, though it is quite possibly a contributing factor, along with changes in exposure (Brun *et al.*, 1997). It is clear, though, that flood, hail and storms have been very costly. With climate change is expected to come more frequent flooding, more droughts and quite possibly more hailstorms and tornadoes, and it is likely that losses will increase in the future.

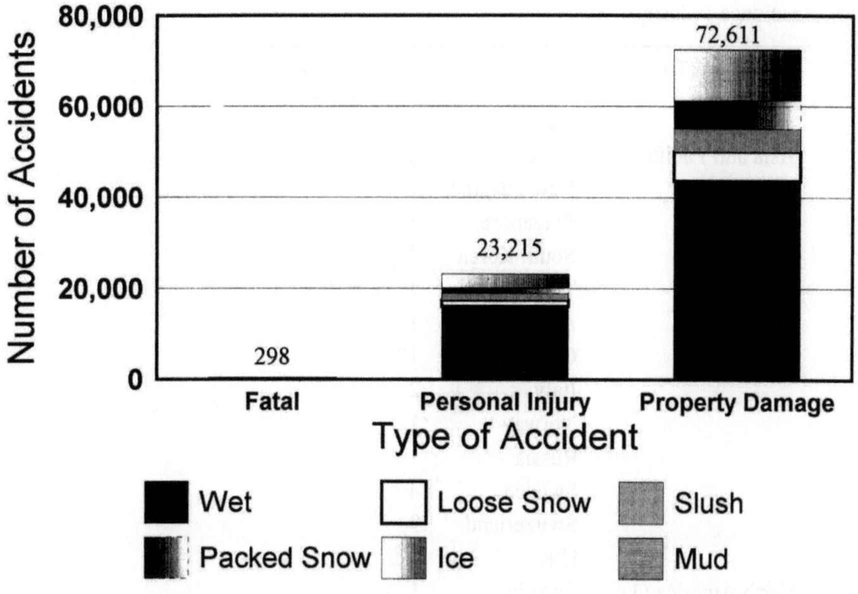

Figure 7. Weather related Ontario road accidents, 1992. Source: Ontario Ministry of Transportation.

4. Reactions in the Insurance Industry to Recent Weather Events

The seriousness of this predicament has not convinced the insurance industry as a whole to advocate mitigation of greenhouse gas emissions. As in other economic sectors, opinion is divided. About 50 insurance companies have signed a declaration with the United National Environment Programme on 'The Insurance Industry and the Environment (Table III)', yet the only companies to call publicly for a reduction in greenhouse gas emissions are the major reinsurers based in Europe (Munich Re, 1996; Swiss Re, 1994).

Although opinion in the insurance industry has not yet coalesced behind a call for emission reduction, it is fully understood that the industry is dangerously overexposed by events such as billion-dollar hurricanes and windstorms (Berz, 1988, 1993; Berz and Conrad, 1994; Insurance Institute for Property Loss reduction 1994; Leggett, 1994; Munich Re, 1990). The 1987 and 1990 storms in western Europe each cost the U.K. property and casualty industry over a billion pounds *in a single day* (Dlugolecki *et al.* 1994). In the United States the wake-up call was hurricane Andrew, followed in 1995 by hurricanes Luis (US$1.5 billion) and Opal (US$2.1 billion) which reinforced the message. On both sides of the Atlantic it is understood that, heavy as these losses had been, they did not represent the worst-case scenario for which the industry is legally and morally obliged to prepare. In Britain it has been calculated that a combination of coastal flooding and windstorm

Table III. National origin of companies which signed the UNEP 'statement on environmental commitment by the Insurance Industry'

Continent (Total)	Country	No. of Signees
Africa (1)	Tanzania	1
Asia and Pacific (9)	Japan	5
	New Zealand	1
	Singapore	1
	South Korea	1
	Thailand	1
Europe (35)	Austria	1
	Germany	13
	Italy	2
	Norway	1
	Russia	1
	Sweden	4
	Switzerland	9
	U.K.	4
North America (1)	Canada	1

damage could easily produce a £5 billion event; flooding in the London area could cost £10 billion. In Florida, an Andrew-force hurricane striking Miami could result in a US$40 billion loss. Whether such extreme events are an anomaly, or part of a trend, the industry must prepare for them.

Immediately following Andrew, it became clear that the industry – globally – did not have the capacity to cover the worst case event. Primary insurance companies responded mainly by taking the traditional steps. Reserves were increased, more reinsurance was purchased where possible; deductibles and premiums were raised – again where possible – and companies that found themselves over-concentrated in geographical high-risk regions reduced their exposure (*National Underwriter*, 25 March 1996.) The fifth largest auto and home insurer in the United States recently announced restrictions on sales in 17 coastal states from Maine to Texas. The move was described by a company spokesman as a 'significant' retrenchment. (*Wall Street Journal*, 10 October 1996).

Even in combination, these measures were not enough to close the gap because the pressure of regulation and the need to maintain a positive public image (and market share), prevent dramatic changes in policy commitments from taking place. Renewed concern with loss prevention brought the industry back to confront the long-standing problem of lack of compliance with building codes. The growing use of predictive computer models provided more realistic and up-to-date estimates of exposure. The Chicago Board of Trade developed markets for catastrophe options in an effort to tap into the liquidity of global capital markets (Hague 1996). A new reinsurance market, backed by Wall Street investment banks, emerged in Bermuda.

Within the industry in the United States, the favoured solution to this difficult situation is to develop a supplementary local reinsurance pool through industry-wide contributions and a public guarantee, such as those being put in place by the Joint Underwriters Association in Florida and the Coastal Market Assistance Program in New York (*National Underwriter*, 18 March 1996). For a fully private sector solution the insurers are asking for tax relief on dedicated funds to be placed in reserve to cover future catastrophes. Twenty to twenty-five years is the kind of time-frame being estimated before such reserves would meet the market's needs.

Even with government support and industry-wide cooperation these proposals can be considered as an adequate response only if these extreme events are viewed as an anomaly, and not as a long term trend associated with an enhanced greenhouse effect. As a reaction to events that have already occurred each the solutions adopted will strengthen the industry's ability to maintain some insurance coverage even in high-risk areas. However, a more proactive campaign to strengthen the demand for mitigative measures (to reduce the probable rate of climate change), is an essential next step. Right now, neither the insurance industry nor governments are prepared for climate change. Nor are they prepared for related societal trends that are increasing the pressure on the insurance industry.

It is extraordinarily difficult for the insurance industry to prepare for climate change; market forces and competition tend to be the driving forces, and the market is reactive in nature, not proactive. Assessing risk using long periods of data with careful consideration of climate variability, cycles and trends, more involvement in adaptation and mitigation measures and research, and risk differentiation according to specific hazards may be helpful approaches.

5. Societal Trends

The rate of global population growth has been falling since the 1960s but it still runs at 1.6 % per annum, adding nearly 100 million people per year, mostly in developing countries. Inevitably this leads to greater population density and hence more people exposed to perils in any particular area. Despite entrenched poverty in some countries and intermittent recession in many others the global economy is still growing, and one result of this growth is that households and companies acquire more goods – especially expensive and fragile consumer durables. To the single radio a Western middle class family might have possessed 50 years ago, we have added two or three automobiles, many radios, several televisions, VCRs, a refrigerator and a freezer, cameras, mobile phones and several computers. This Western pattern of material acquisition is now spreading rapidly to the emerging economies of Latin America, East and South Asia and to the economies in transition in central Europe.

Two further societal trends are of importance to our argument. Population growth and continued growth of material possessions are accompanied by urbanisation. The urbanisation trend in the first industrial countries (now at the 80% mark), is losing momentum, but in the emerging economies there is still a long way to go. Population growth and urbanisation combine to produce ever-larger settlements, with 25 having a projected population greater than 10 million by the Year 2000, of which 20 will be found in what used to be called the developing world. Air quality and water quality are now serious problems in many major cities, especially as the conjunction of economic growth produces more and more people dependent on private automobiles for their daily travel, further contributing to both traffic congestion and urban sprawl. When a storm strikes, a greater percentage of people are exposed in their cars on highways because that is where they spend more and more of their time.

One last trend – which we will call 'coastalisation' – is further concentrating people in high risk areas as they move to the coast in search of economic opportunity, recreation or retirement. As agriculture reduces its need for labour and as the coal mines lose their attraction for manufacturing enterprises, there are fewer and fewer reasons for people to inhabit the interiors of continents. The coastal movement began in the United States at the early part of this century and the same trend is now gathering momentum in China. Thus, it is a phenomenon which is prevalent in both the richest countries and the poorest. Coasts are areas of relatively high risk for natural hazards because of their vulnerability to storms, tsunamis and flooding (as well as earthquakes). If sea-level rise continues (as is predicted by the models of the enhanced greenhouse effect) this will further increase the risk.

Canada is something of an exception to this global trend in that the Atlantic coast is not a magnet for population. In terms of economic opportunity the draw in central and eastern Canada is the St. Lawrence Valley, focused on the Greater Toronto Area, and this is not an area which is particularly susceptible to natural hazards. However, the same dangers of concentration of insured wealth exist, given the high rate or urbanisation and the concentration of Canada's population in just three urban areas. The West Coast is a classic case of concentrated coastal risk, where the Greater Vancouver Area is situated in an active fault zone, and is at risk to tsunamis. In search of recreation and retirement, Canadians conform to the global trend in seeking coastal zones with a warmer climate, especially Florida.

These trends are producing a greater number of people with more and more goods living in coastal cities and other built-up areas, and to a large extent account for the fact that billion-dollar losses have become commonplace in recent years (*Reinsurance*, 1996). To some extent this trend could be reversed through land-use planning (by not allowing people to live in harm's way), or by ensuring that people assume their own risk. Although a great deal has been written on the problems of managing these huge urban agglomerations, very little has been produced specifically on the management of urban risks. The professions of 'risk management and insurance' and 'urban management' have yet to form any significant partnerships.

6. Climate Change and Health

Human health must also be considered when examining the climate change scenario as it relates to insured risks, specifically the life and health insurance industry. There are a variety of expected changes – some bring a fairly direct causality, others are more indirect and speculative; they include the positive and the negative, with the negative preponderant. As noted earlier, many scientists expect that the enhanced greenhouse effect will produce more extreme events such as droughts, floods, and heat waves. All of these can increase mortality and morbidity, as was seen dramatically in the 1995 heat wave in North America to which 465 excess mortalities were identified in the Chicago area alone, mostly for old people who died from heat stress as a power failure cut off their electric fans and air conditioners (NOAA, 1995).

Although extreme events such as droughts, floods and heat waves will produce the most immediate impacts on human health, it is likely that some of the less dramatic changes that are expected to accompany climate change, will – in the longer run – have even more serious consequences. For example, medical scientists are concerned about the extension of vector-born diseases such as malaria and yellow fever as the climate warms and creates favourable conditions for the hosts over a wider area (Haines *et al.*, 1993). At the end of the last century malaria was the number one cause of mortality in the United States and it could return as a significant health factor. Warmer temperatures would also be favourable to cholera, especially in areas with poor water quality. An additional factor which would increase the health risk is the increased mobility of the global population, as has been seen with the resurgence of tuberculosis in Europe. Mobility could increase further in response to environmental stress caused by greater climate variability and coastal flooding. Human health would also be susceptible to increased exposure to ultra-violet radiation due to the depletion of stratospheric ozone.

The most serious health implications of climate change may be the most indirect, through crop failures and livestock losses, the decline of fisheries due to overfishing and the impact of ultra-violet radiation on phytoplankton, and the decreasing availability of clean water. The implications of many of these stresses are indirect for Canada. However, as a relatively low-risk region it would become a magnet for people moving from less fortunate places.

7. Issues for the Insurance Industry

It cannot be denied that even *the possibility* of climate change injects a great deal of uncertainty into nearly every aspect of human society. Ironically, much of the apparent progress of human society – in material terms at least – has been won by reducing the risks to which humans are exposed, through activities such as crop storage, education, medical services, safe water supply, weather forecasting, and the evolution of the insurance industry itself. Yet, we are now willing to raise that level of risk by changing the composition of the atmosphere, merely for the convenience of enjoying cheap fossil fuels in the short term.

A further question is raised by humankind's deliberate alteration of the composition of the atmosphere if this action results in global warming, sea-level rise and more extreme events. The question is: if these events are deliberately induced, can they still be seen as 'acts of God', such an act being viewed as 'a direct, violent sudden act of nature that could not have been foreseen, or if foreseen, its effect could not have been prevented' (The Insurance Institute of Ontario, 1991, 3)? In other words, if we create these conditions, are the ensuing risks insurable at all? A recent ruling in a California court found in favour of the insurers of a client (who had polluted the groundwater around several of its factories), on the grounds that the client 'should have known' that these actions would lead to contamination (*National Underwriter*, 1 April 1996). This subjective criteria opens up some very broad questions about the deliberate emission of greenhouse gases – the largest single act of pollution that human beings have contrived so far – at a time when the global scientific community has concluded that 'potentially serious changes have been identified, including an increase in some regions of extreme high temperature events, floods, and droughts, with resultant consequences for fires, pest outbreaks and ecosystem composition, structure, and functioning, including primary productivity' (IPCC, 1995, 4).

As the implications of global warming become clearer, the victims of climate change and variability will find it difficult to put claims to their insurer on the assumption that the storms are both sudden and unpredictable. Indeed the mathematician, Fourier, predicted this outcome in the 1820s, while the physicist Arrhenius published the first paper on global warming in 1893, more than 100 years ago.

Climate change, and the extreme weather events that it may bring, pose several questions for the insurance industry, both short term and long term. The immediate issues relate to the need to build up reserves as quickly as possible to cover the maximum possible loss, while simultaneously taking all steps to reduce exposure and mitigate losses. The constraints of regulation, social responsibility and competition within the industry make it impossible to simply withdraw coverage from those places where the risks appear to have increased, such as hurricane-prone coastal regions, which is why voluntary pools of regional reinsurance are being established for Florida and Long Island.

Both temporal and spatial elements of the insurance industry make it difficult to move in the right direction. The highly competitive nature of the industry drives 'cycles' of underwriting which force premiums down to the point where companies exit the industry as losses mount, and then premiums rise again to cover the losses. In the 1980s aggressive competition for market share was driven by the profits to be made by investing the industry's reserve funds. Underwriting losses were routinely covered by investment profits, thus creating a situation which negated the fundamental objective of underwriting, which is to write policies which earn a profit (Holton, 1987). In this type of situation it is difficult to assemble coalitions of insurers who agree to share the less attractive risks in a regional pool. This could be a situation where the virtues of competition serve neither the shareholders nor the public interest.

The insurance industry is based on the randomness of events in both time and space. Insurers normally do not insure spatially concentrated risks (like dwellings on flood plains), because the premiums would be unattractive to potential clients. The ideal market is one based on a ubiquitous risk, potentially devastating, and of very low probability. In such a situation many people will want to buy insurance, but very few will ever make a claim. Thus the geographical specificity of the risks posed by climate change and extreme events is a very uncomfortable problem for the industry to face. Can homeowners in Alberta be expected to pay higher premiums to cover the expected cost of increased hurricane damage in Nova Scotia, as Nova Scotians might currently cover tornado risk for Ontario and hailstorm risk for Alberta?

As the impacts of climate change become more obvious to the public and to politicians, the deeper implications of responsibility for these changes will have to be addressed. Some examples of the dilemmas this will pose are predictable from current issues such as the class suits being brought against the tobacco industry. Were smokers 20 years ago supposed to know that smoking was addictive and life-threatening? Today, are users of fossil fuels (meaning all of us in the richer countries and many in the poorer countries) supposed to know that our actions are changing the composition of the atmosphere, with unpredictable – but probably dire – results?

8. Solutions

One option for the industry would be to assist in educating the public and the politicians to understand the need to reduce greenhouse gas emissions as speedily as possible. Other efforts should go into monitoring and research activities that can enable us to adapt to the climate changes to which our actions have already committed us. The first issue for all players in the industry (and its regulators) to face is that climate change will increase the level of uncertainty. Accepting this proposition will require traditional responses such as reserve accumulation, loss reduction, and so on. It also points toward the adoption of a much more pro-active stance on mitigation through the reduction of the use of fossil fuels, even though this is likely to result in a head-on confrontation with the fossil fuel lobby and its supporters. The alternative is to join the rest of society and 'wait-and-see' if the current scientific consensus is correct in assuming that the warming trend is detectable and that it will most likely result in more droughts, floods and extreme weather events.

References

Agee, M.: 1991, Trends in cyclone and anticyclone frequency and comparison with periods of warming and cooling over the northern hemisphere, *J. Climate* **4**, 263–267.

Arrhenius, E. and Waltz, T.: 1990, Minimizing the greenhouse effect, in: Alcira Kreimer and Mohan Munasinghe (eds.), *Managing Natural Disasters and the Environment*, World Bank, June 27–28, Washington, D.C.

Balling, R. C. and Lawson, M. P.: 1982, Twentieth century changes in winter climatic regions, *Climatic Change* **4**, 57–69.

Barrow, E. M. and Hulme, M.: 1996, Changing probabilities of daily temperature extremes in the UK related to future global warming and changes in climate variability, *Climate Res.* **6**, 21–31.

Beer, T. and Williams, A.: 1995, Estimating Australian forest fire danger under conditions of doubled carbon dioxide concentrations, *Climatic Change* **29**, 169–188.

Berz, G.: 1988, Climatic change: Impact on international reinsurance, in: G. I. Pearman (ed.), *Greenhouse. Preparing for Climate Change*, CIRO, pp. 579–587.

Berz, G.: 1993, Global warming and the insurance industry, *Interdisciplinary Sci. Rev.* **18**(2), 120–125.

Berz, G. and Conrad, K.: 1994, Stormy weather: the mounting windstorm risk and consequences for the insurance industry, *Ecodecision*, April, 65–68.

Broccoli, A. J. and Manabe, S.: 1990, Can existing climate models be used to study anthropogenic changes in tropical cyclone climate, *Geophys. Res. Lett.* **17**(11), 1917–1920.

Broecker, W. S.: 1995, Chaotic climate, *Scientific American* November, 62–68.

Brun, S. E., Etkin, D., Law, D. G., Wallace, L., and White, R.: 1997, Coping with natural hazards in Canada: Scientific, government and insurance industry perspectives, Insurers' Advisory Org. Inc., Toronto, Canada, 217 pp.

Carnell, R. E., Senior, C. A., and Mitchell, J. F. B.: 1996, An assessment of measure of storminess: Simulated changes in northern hemisphere winter due to increasing CO_2, *Climate Dynamics* **12**, 467–476.

Changnon, S. A. and Changnon, J. M.: 1992, Temporal fluctuations in weather disasters: 1950–1989, *Climatic Change* **22**, 191–208.

Cubash, U., Waszkewitz, J., Hegerl, G., and Perlwitz, J.: 1995, Regional climate changes as simulated in time slice experiments. MPI Report 153, *Climatic Change* **31**, 273–304.

Dlugolecki, A., Elvy, C., Kirby G., Salthouse, R., Turner, S., Witt, D., Martin, R., Toomer, C., Secrett, B., Palutikoff, J., and Clement, D.: 1994, Task force on 'The impact of changing weather patterns on property insurance', Chartered Insurance Institute, London, May, 87 pp.

Emmanuel, K. A.: 1987, The dependence of hurricane intensity on climate, *Nature* **326**, 483–484.

Emmanuel, K. A.: 1995, Comments on 'Global climate change and tropical cyclones': Part 1, *BAMS* **76**(11), 2241–2244.

Environment Canada: 1988, *Conference Statement* from 'The Changing Atmosphere: Implications for Global Security', Toronto, June 27–30.

Environment Canada: 1996, *Summer 1996 – The Season That Almost Wasn't*, Accessed from website ⟨http.//www.on.doe.ca./comm⟩, September 18.

Evans, J. L.: 1993, Sensitivity of tropical cyclone intensity to sea surface temperature, *J. Climate* **6**, 1133–1140.

Etkin, D. A.: 1995, Beyond the year 2000, more tornadoes in western canada? Implications from the historical record, *Natural Hazards* **12**, 19–27.

Gordon, H. B., Whetton, P. H., Pittock. A. B., Fowler, A. M., and Haylock, M. R.: 1992, Simulated changes in daily rainfall intensity due to the enhanced greenhouse effect: Implications for extreme rainfall events, *Climate Dynamics* **8**, 83–102.

Griffiths, D. J., Colquhoun, J. R., Batt, K. L., and Casinadar, T. R.: 1993, Severe thunderstorms in New South Wales: Climatology and means of assessing the impact of climate change, *Climatic Change* **25**, 369–388.

Haarsma, R. J., Mitchell. J. F. B., and Senior, C. A.: 1993, Tropical disturbances in a GCM, *Climate Dynamics* **8**, 247–257.

Hague, K.: 1996, Derivatives: bridge to the capital markets *Canadian Underwriter*, April 1996, 30–35.

Haines, A., Epstein, P., and McMichael, A.: 1993, Global health watch: Monitoring impacts of environmental change, *The Lancet* **342**, 1454–69.

Hall, N. M. J., Hoskins, B. J., Valdes, P. J., and Senior, C. A.: 1994, Storm tracks in a high resolution GCM with doubled CO2. *Quart. J. Royal Meteorol. Soc.* **120**, 1209–1230.

Hansen, J., Fung, I., Lacis, A., Rind, D., Lebedoff, S., Ruedy, R., and Russel, G.: 1988, Global climate changes as forecast by GISS's three-dimensional model, *J. Geophys. Res.* **93**, 9341–9364.

Held, I. M.: 1993, Large-scale dynamics and global warming, *BAMS* **74**(2), 228–241.

Hennessy, K. J. and Pittock, A. B.: 1995, Greenhouse warming and threshold temperature events in Victoria, Australia, *Int. J. Climatology* **15**, 591–612.

Hogg, W. D.: 1996, Cycles and trends in time series of Canadian extreme rainfall, unpublished manuscript, Environment Canada, Downsview.

Holton, R. B.: 1987, *Underwriting Principles and Practices*, 3rd edn., The National Underwriter Company, Hoboken, N.J., USA.

Hughes, M. K and Brown, P. M.: 1992, Drought frequency in Central California since 101 BC. Recorded in giant sequoia tree rings, *Climate Dynamics* **6**, 161–167.

Hulme, M.: 1992, Rainfall changes in Africa: 1931–1960 to 1961–1990. *Int. J. Climatology* **12**, 685–690.

Idso, S. B., Balling, R. C., and Cerveny, R. S.: 1990, carbon dioxide and hurricanes: Implications of northern hemispheric warming for Atlantic/Caribbean storms, *Meteorol. Atmos. Phys.* **42**, 259–263.

Insurance Institute for Property Loss Reduction: 1994, *Understanding the Wind Peril*, Boston

Insurance Institute of Ontario: 1991, *Dictionary of Insurance*, Toronto.

IPCC: 1995, *Climate Change 1995: Impacts, Adaptations, and Mitigation*, Summary for Policy makers. Contributions of Working Group II to the Second Assessment Report of the Intergovernmental Panel on Climate Change, WMO and UNEP.

IPCC: 1995, Climate Change 1995: in: J. T. Houghton, L. G. Meira Filho, B. A. Callander, N. Harris, A. Kattenberg, and K. Maskell (eds), *The Science of Climate Change*, Cambridge University Press.

Karl, T. R., Knight, R. W., Easterling, D. R., and Quayle, R. G.: 1995, Indices of climate change for the United States, *Bull. Am. Meteorol. Soc.* **77**(2), 279–292.

Karl, T. R., Knight, R. W., and Plummer, N.: 1995, Trends in high frequency climate variability in the twentieth century, *Nature* **377**, 217–220.

Khandekar, M. L. and Swail, V. R.: 1995, Storm waves in Canadian waters: A major marine hazard, *Atmos.-Ocean* **33**(2), 303–328.

Karl, T. R., Knight, R. W., Easterling, D. R., and Quayle, R. G.: 1995, Trends in U.S. climate during the twentieth century, *Consequences* (Saginaw Valley State University), **1**(1), 2–12.

Katz, R. W. and Brown, B. G.: 1992, Extreme events in a changing climate: Variability is more important than averages, *Climatic Change* **21**, 289–302.

Kovacs, P.: 1997, Insurance Bureau of Canada.

Kurz, W. A. and Apps, M. J.: 1996, Retrospective assessment of carbon flows in Canadian boreal forests, *Forest Ecosystems, Forest Management and the Global Carbon Cycle*, NATO ASI series, vol. I, **40**, 173–182.

Lambert, S. J.: 1995, The effect of enhanced greenhouse warming on winter cyclone frequencies and strengths, *J. Climate* **8**, 1447–1452.

Lambert, S. J.: 1996, Intense extratropical northern hemisphere winter cyclone events: 1899–1991, *J. Geophys. Res.* **101**(D16) **21**, 319–21,325.

Landsea, C. W., Nicholls, N., Gray, W. M., and Avila, L. A.: 1996, Downward trends in the frequency of intense Atlantic hurricanes during the past five decades, *Geophys. Res. Lett.* **23**, 1697–1700.

Lawford, R. G., Prowse, T. D., Hogg. W. D., Warkentin, A. A., and Pilon, P. J.: 1995, Hydrometeorlogical aspects of flood hazards in Canada, *Atmos. Ocean* **33**(2), 303–328.

Leathers, D. J. and Ellis, A. W.: 1996, Synoptic mechanisms associated with snowfall increases to the Lee of Lakes Erie and Ontario, *Internat. J. Climatol.* **16**, 1117–1135.

Leggett, J.: 1994, Climate change and the financial sector, Lecture delivered to the Society of Fellows, Chartered Insurance Institute, London, November 1.

Lighthill, J., Holland, G., Gray, W., Landsea, C., Craig, G., Evans, J., Kurihara, Y.m and Guard, C.: 1994, Global climate change and tropical cyclones, *BAMS* **75**(11), 2147–2157.

Manabe, S and Stouffer, R. J.: 1988, Two stable equilibria in a coupled ocean-atmosphere model, *J. Climate* **1**, 841–866.

Maybank, J., Bonsal, B., Jones, K., Lawford, R., O'Brian, E. G., Ripley, E. A., and Wheaton, E.: 1995, Drought as a natural disaster, *Atmos. Ocean* **33**(2), 195–222.

Mearns, L. O., Katz, R. W., and Schneider, S. H.: 1984, Extreme high-temperature events: Changes in their probabilities with changes in mean temperature, *J. Climate Appl. Meteorol* **23**, 1601–1613.

Mearns, L. O., Giorgi, F., McDaniel, L., and Shields, C.: 1995, Analysis of variability and diurnal range of daily temperature in a nested regional climate model: Comparison with observations and doubles CO_2 results, *Climate Dynamics* **11**, 193–209.

Meehl, G. A.: 1993, Changes of variability in a climate with increased CO_2: El Nino-Southern oscillation and the Asian summer monsoon, In *Tol* (1993).

Mitchell, J. F. B. and Ingram, W. J.: 1990, On CO_2 and climate. Mechanisms of changes in cloud, *J. Climate* **5**, 5–21.

Munich Re: 1990, *Windstorm – New Dimensions of a Natural Hazard*, Munich.

Munich Re: 1996, *Annual Review of Natural Catastrophes 1995*. Extract from the Preprint, 7.

National Science and Technology Council: 1996, Natural Disaster Reduction: A Plan for the Nation. The White House, Washington, USA.

National Underwriter, March 18, 1996, Voluntary market facility set up for New York coastal risks, **40**, 41.

National Underwriter, March 25th, 1996, Florida property insurers balk at sweeping package of reforms, p. 33.

National Underwriter: April 1st 1996, California jury hands Lloyd's pollution win.

NOAA, 1995: July 1995 Heat Wave, Natural Disaster Survey Report, National Weather Service, Silver Spring, Maryland, 53 pp.

Noda, A. and Tokioka, T.: 1989, The effect of doubling CO_2 concentration on convective and non-convective precipitation in a general circulation model coupled with a simple mixed layer ocean, *J. Met. Soc. Japan* **67**, 95–110.

Oladipo, E. O.: 1993, Drought in Northern Nigeria: An indication of abrupt climatic change, *Weather and Climate* **13**, 34–39.

Pang, A.: 1997, AON Re.

Parey, S.: 1994, Simulations de trente ans $1 \times CO_2$, $2 \times CO_2$, $3 \times CO_2$ avec le modèle du LMD (64 $\times$ 50 $\times$ 11) premiers résultants, EDF, Direction des études et recherches, HE-33/94/008.

Pielke Jr., R. A. and Landsea, C. W.: 1997, Normalized hurricane damages in the U.S.: 19250–1995, Draft paper.
Http://www.dir.ucar.edu/esig/HP_roger/hurr_norm.htm.

Rowntree, P.: 1993, Workshop on 'Socio-Economic and Policy Aspects of Change of Incidence and Intensity of Extreme Weather Events', Institute for Environmental Studies, W93/15, Free University, Amsterdam, June 24–25, 1993.

Price, C. and Rind, D.: 1993, Lightning fires in a $2 \times CO_2$ world, in *Proc. 12th Conference on Fire and Forest Meteorology*, Oct. 26–28, Jekyll Is., Georgia, pp. 77–84.

Reinsurance: 1996, Megacities under threat, **26**(10), 28–29.

Rind, D., Goldberg, R., and Ruedy, R.: 1989, Change in climate variability in the 21st century, *Climate Change* **14**, 5–38.

Ross, A.: 1996, Climatic change and its impact on the Canadian insurance industry, seminar delivered at the Institute for Environmental Studies, University of Toronto, February 29th.

Ross, R. J. and Elliot, W. P.: 1996, Tropospheric water vapor climatology and trends over North America: 1973–93, *J. Climate* **9**, 3561–3574.

Ryan, B. F., Watterson, I G., and Evans, J. L.: 1992, Tropical cyclone frequencies inferred from Gary's yearly genesis parameter: Validation of GCM tropical climates, *Geophys. Res. Lett.* **19**(18), 1831–1834.

Singer, S. F., Boe, B. A., Decker, F. W., Frank, N., Gold, T., Gray, W., Linden, H., Lindzen, R., Michaels, P. J., Nierenberg, W. A., Porch, W., and Stevenson, R.: 1997, Comments on 'Open Letter to Ben Santer', *BAMS* **78**(1), 81–82.

Skaggs, R. H., Baker, D. G., and Ruschy, D. L.: 1995, Interannual variability characteristics of the Eastern Minnesota (USA) temperature record: Implications for climate change studies, *Climate Res.* **5**, 223–227.

Slivitzky, M. and Morin, G.: 1996, Impacts of climatic changes on the hydrological regime: The Moisie River case revisited, *CMOS Bull.* **24**(4), 77–81.

Smith, D. I.: 1993, Greenhouse climatic change and flood damages, the implications, *Climatic Change* **25**, 319–333.

Stein, O. and Hense, A.: 1994, A reconstructed time series of the number of extreme low pressure events since 1880, *Meteorol. Zeit.* **3**, 43–46.

Street, R. B.: 1989, Climate change and forest fires in Ontario, in *Proc., 10th Conference on Fire and Forest Meteorology*, Ottawa, Ontario, pp. 177–181.

Swiss Re: 1994, *Global Warming: Element of Risk*, Zurich.

Wall Street Journal: October 10th 1996, 'Nationwide To Unveil Rules To Cut Risks After Hurricanes'

Tol, R. S. J. (ed.): 1993, Socio-economic and policy aspects of changes in the incidence and intensity of extreme weather events, institute for environmental studies, W93/15, Free University, Amsterdam, June 24–25.

Vance, R. E.: 1991, A Paleo-botanical study of holocene drought frequency in Southern Alberta, PhD Thesis, University of British Columbia, 180 pp.

Wetherald, R. T. and Manabe, S.: 1995, The mechanisms of summer dryness induced by greenhouse warming, *J. Climate* **8**, 3096–3108.

Whetton, P. H. J., Fowler, A. M., Haylock, M. R., and Pittock, A. B.: 1993, Implications of climate change due to the enhanced greenhouse effect on floods and droughts in Australia, *Climatic Change* **25**, 289–317.

Wigley, T. M. L.: 1988, The effect of changing climate on the frequency of absolute extreme events, *Climate Monitor* **17**, 44–55.

Wilson, C. A. and Mitchell, J. F. B.: 1987, Simulated climate and CO2 induced climate change over Western Europe, *Climatic Change* **10**, 11–42.

Zwiers, F. W.: 1994, Changes in screen temperature extremes under a doubling of CO2, in G. J. Boer (ed.), *Research Activities in Atmospheric and Oceanic Modelling*, 7.44–7.46, CAS/JSC Working Group Numerical Experimentation, Report 19, Feb., WMO/TD-No 592.

Natural Hazards **16:** 165–179, 1997.

Problems for Probabilistic Seismic Hazard Analysis

D. H. SPEIDEL and P. H. MATTSON
Department of Geology, Queens College and the Graduate School, City University of New York, 65-30 Kissena Blvd., Flushing, NY 11367, U.S.A.

(Received: 4 December 1996; in final form: 9 September 1997)

Abstract. The use of *b*-values derived from the Gutenberg–Richter relationship as a phenomenological base for developing probabilistic seismic hazard analyses (PSHA) has been questioned for years. The relationship is still used because political demands require something for PSHA, one variable is easy to deal with, and no persuasive alternative has come forward. Using cumulative distribution probability plots, it can be shown that seismic magnitude-frequency data can be well described as one or more populations, each of which is normally distributed with respect to magnitude. This holds true for large earthquakes when sorted by mechanism, for earthquakes >400 km deep, for the general USGS NEIC catalog, for the Harvard CMT catalog, for the CERI catalog of the New Madrid Zone, and for a Scandinavian catalog. In all instances, multiple normal populations provide a better fit to the data than does the Gutenberg–Richter relationship. Use of these multiple populations in PSHA emphasizes that the scientifically sound limits of magnitude projection are within the 4σ limit of the largest populations. Such graphs may make it easier to resist political requirement to extrapolate into scientifically unsound regions.

Key words: earthquake prediction, magnitude-frequency relationship, PSHA.

1. Introduction

Seismic hazard is the probability of the occurrence of such earthquake-related phenomena as ground shaking, fault movement, soil liquefaction, or landslides (Reiter, 1990). One way to estimate that hazard is through Probabilistic Seismic Hazard Analysis (PSHA). For PSHA three types of input are required: (i) earthquake location(s) in space and the resulting epicentral distances to the area of interest; (ii) a magnitude distribution and rate of occurrence for each source identified in (i); and (iii) ground motion estimations associated with a particular earthquake from a particular source with a particular magnitude (NRC, 1988). While PSHA does not require a particular distribution for (ii) (see, for example, NRC, 1988, Figure 2.1), most common treatment (Reiter, 1990, Figure 10.2) assume that the seismicity follows the Gutenberg–Richter relationship (Gutenberg and Richter, 1949). This relationship is most commonly written (Lomnitz, 1994, p. 161) as

$$\log N = a - bM, \tag{1}$$

where N is the number of earthquakes for either a magnitude increment or cumulative total; M is the size of the earthquake in any linear magnitude or intensity scale or the log of moment (Krinitzsky, 1993) and a and b are constants. N can also be for particular regions or specific time intervals. In this formulation, a is the log of the number of events ($M > M_0$) of a particular area (= log N when M = 0) and b is a measure of the mean magnitude (Lomnitz, 1974, p. 92). Graphically determined from the slope of a frequency magnitude plot, b is dependent on the relative proportions of small, medium, and large shocks. The compilation of small and medium events in the historical or instrumental record allows estimation of a b-value which, when projected to higher magnitudes is (incorrectly) used to predict recurrence intervals for less-common, more damaging earthquakes. The sensitivity of the postulated recurrence interval for high magnitudes to relatively small changes in b-value, make estimation of b important (Reiter, 1990). Thus, any challenge to the validity of Equation (1) has a direct bearing on PSHA as it usually is employed.

The relationship between magnitude and frequency is one of the most investigated ones in the seismological literature (Båth, 1981). Many papers in addition to those of the authors cited later argue that the Gutenberg–Richter relationship is valid only with a range of magnitudes or does not hold up at all (Aki, 1987; Frohlich and Davis, 1993; Main, 1987; Pacheco *et al.*, 1992; Papadopoulos *et al.*, 1993; Purcaru and Zorilescu, 1971; Ranalli, 1975; Rydelic and Sacks, 1989; Sacuiu and Zorilescu, 1970; Scholz, 991; Taylor *et al.*, 1990; Trifu *et al.*, 1993). Sornette *et al.* (1996) point out that there is no theoretical basis for the power law. They state that the distribution that holds for small earthquakes can not be extended to infinity because it would require an infinitely large release of energy. As an alternative, Sornett *et al.* (1996) argue based on Zipf's law of least effort (Zipf, 1949) that several populations exist – one at small and medium populations and the other at high magnitudes. All the same, recent papers assume the validity of the Gutenberg–Richter relationship (El-Sayed and Wahlström, 1996; Espinosa, 1996).

The geologic record does not agree with what would be predicted using b-values (Krinitzsky, 1993; Schwartz and Coppersmith, 1984) as illustrated in Figure 1. Schwartz and Coppersmith (1984) proposed the concept of the 'characteristic earthquake', a maximum earthquake that occurs repeatedly on a particular fault or segment with greater frequency than that predicted through b value projections generated from the historical and/or instrumental record of seismicity – a concept that has generated considerable discussion (Wesnousky, 1994; Kagan, 1994).

For many seismologists, fault size is directly related to earthquake magnitude distribution (Bonilla *et al.*, 1984; Wells and Coppersmith, 1994) with straight line regression equations developed for magnitude vs. surface rupture length. Fault length is thus related to magnitude of the characteristic earthquake. Nicol *et al.* (1996) point out that fault size distributions do not follow a power law over a large scale range but possibly follow power laws on smaller scales, each set of which has a different exponent. Whether the Gutenberg–Richter, the characteristic, or some other distribution is an appropriate description for earthquake magnitude is especially important in the case of very large earthquakes (Hough, 1996; Jackson, 1996).

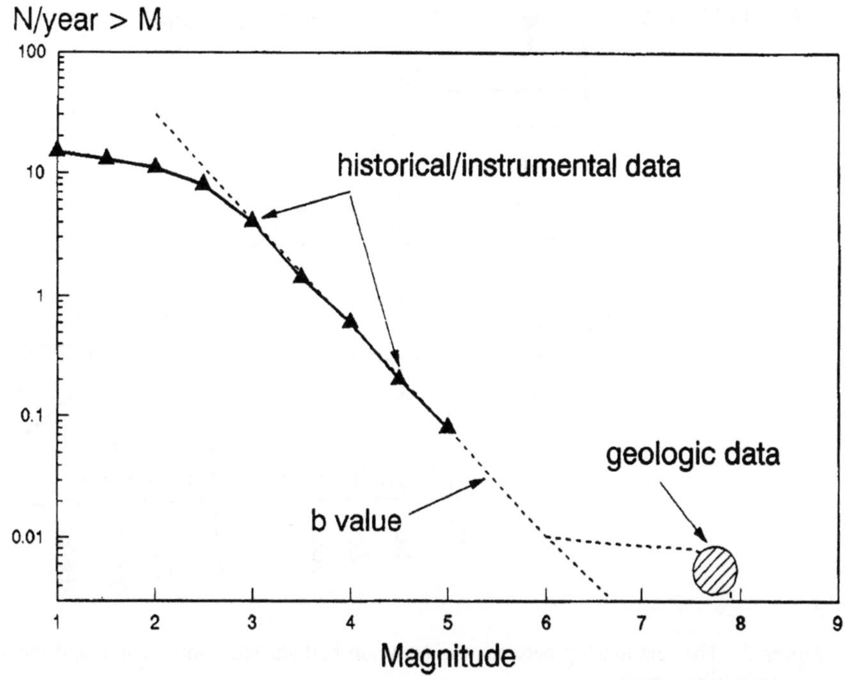

Figure 1. Recurrence relationship difference between Gutenberg–Richter prediction (*b*-value) and the characteristic earthquake (geologic data). After Schwartz and Coppersmith, 1984.

Lomnitz (1974, 1994) argues that the distribution of earthquake energies should be lognormal (and hence earthquake magnitudes should be normal). The exponential Gutenberg–Richter (GR) function fits the tail of a normal distribution. For risk applications, Lomnitz (1974, p. 95) prefers the exponential form rather than the 'normal form of magnitude distribution because a single-parameter distribution affords a more economical description of seismicity without a loss of generality, in the range of large magnitudes'. However, the use of the economic and simpler single parameter GR relationship as a reasonable approximation at high magnitudes for a lognormal distribution of earthquake energies is not valid if multiple populations exist. If such populations can be shown to fit the empirical record of earthquake activity for a particular region in a particular time, the behavior of lognormal distributions (Aitchison and Brown, 1957) allows extension of lognormal distributions or mixtures of such distributions to the whole distribution. This argument is based on the theory of breakage (Aitchison and Brown, 1957) that Lomnitz (1974) uses to argue that stress drop, fault area, and by extension, fault length can all be described as lognormal distributions. Previous work by the authors have presented a variety of examples where earthquake magnitudes can be well described as mixtures of normal distributions (Mattson and Speidel, 1992, 1994, 1996; Speidel, 1996; Speidel and Mattson, 1993a, 1993b, 1994; Speidel *et al.*, 1996).

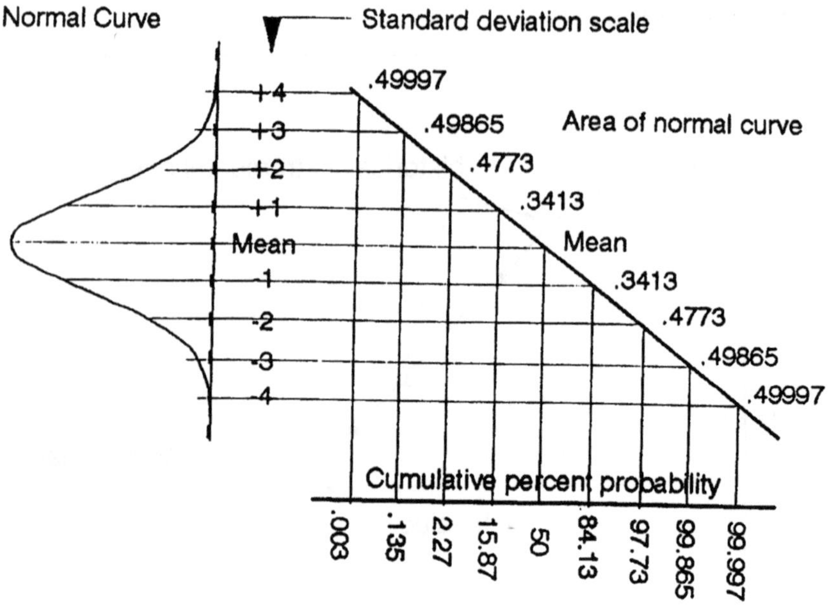

Figure 2. The relationship between the common bell-shaped normal curve and the straight line probability graph.

2. Examples of Normally Distributed Earthquake Populations

As illustrated in Figure 2, any normally distributed population can be represented as a straight line in a probability plot when the abscissa scale is in standard deviation units, that is, as the cumulative distribution function (Titterington *et al.*, 1985). If the data on such a probability graph fit a straight line, the population may be treated as if it followed a normal distribution with the slope of the line a measure of the variance. Speidel and Mattson (1993b) illustrate that such a straight line distribution fits the Giardini Catalog (Giardini, 1988). Thus magnitudes of over 4600 deep (>350 km) worldwide earthquakes could be described as normally distributed, with a mean and standard deviation of 4.65 ± 0.45. Because deep earthquakes are presumed to have a common (although not agreed upon) cause (Frohlich, 1989; Green, 1993), Speidel and Mattson (1993b) argued that earthquake populations are thus associated with mechanisms rather than regions.

Mixtures of populations can be analyzed to determine the parameters of that mixture, i.e., number of populations and mean and variance for each. Titterington *et al.* (1985) illustrate various methods based on the cumulative distribution function. One such is a graphical technique, expanded considerably by Sinclair (1974). Figure 3 illustrates a data set from the northern Europe earthquake catalog of Ahjos and Uski (1992). The 780 earthquakes that occurred north of latitude 70° N fit a pattern that would be expected if two normal distributions were present (Speidel *et al.*, 1996b). The populations are 3.45 ± 0.65 and 1.34 ± 0.5 in the proportions

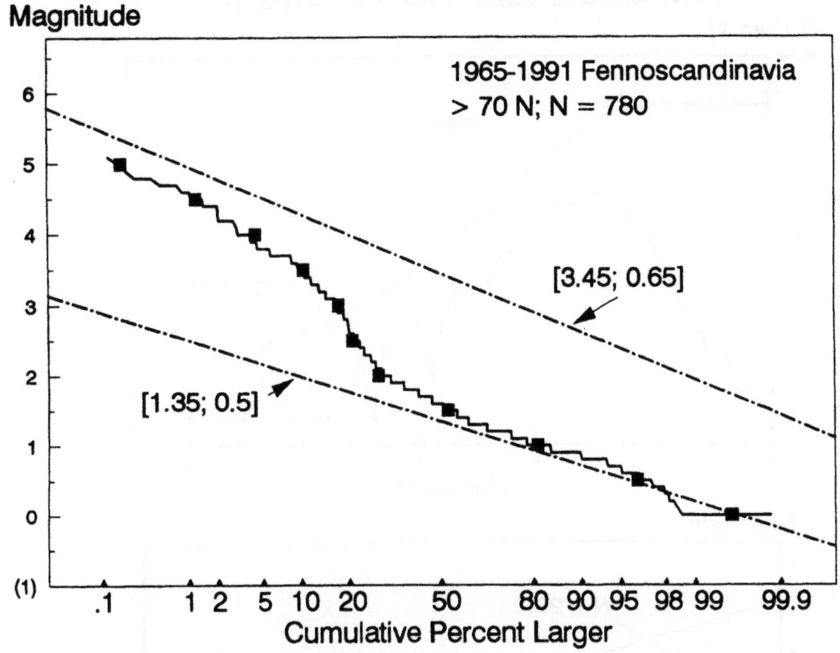

Figure 3. An example of the aggregrate probability pattern or cumulative distribution function produced by adding two normal populations. The combination of 22% of 3.45 ± 0.65 and 78% of 1.35 ± 0.5 produces the filled squares.

78:22. The data are given by the short lines and the two population model is given by the filled squares.

The New Madrid Zone (Mattson and Speidel, 1994; Speidel and Mattson, 1995) illustrates the superposition of several normal populations. Figure 4a is the cumulative frequency-magnitude plot (filled circles) and a frequency plot (filled triangles) for 2852 New Madrid Zone earthquakes between 29 June 1974 and 26 June 1991. It is clear that the data are a poor fit for a straight line exponential function. A plot of the cumulative distribution function on normal probability paper is given in Figure 4b by the heavy line with filled triangles associated with it. Three normal populations were deduced. The 1.6 ± 0.3 population represents 78.2% of the total sample and is associated with thrust activity in the central region. The 1.9 ± 0.3 population represents 21.3% of the activity and appears to be related to strike slip movement along vertical faults. The 4.1 ± 0.6 population, although less than 0.5% of the activity, clearly is necessary to fit the distribution. Unfortunately, the highest magnitude population could not be associated with a particular mechanism (Speidel and Mattson, 1995). The distributions of each subsample is shown in Figure 4c.

Note that the probability of a magnitude 6 occurrence is less than 0.135% ($+3\sigma$) for each incident of the 4.1 population and is effectively zero for the other populations (Figure 4b). This emphasizes the dependence of this type of analyses

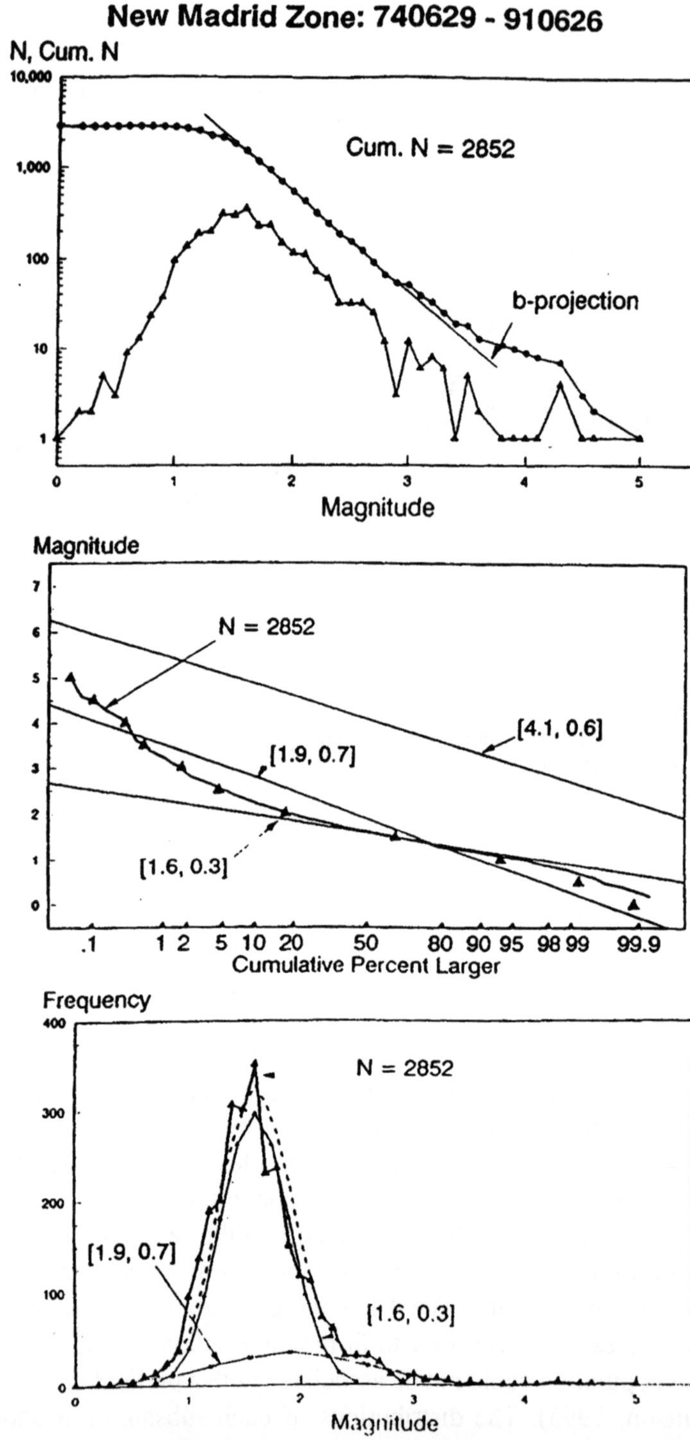

Figure 4. An example of the application of probability graphs for overlapping populations from the New Madrid Zone. (a) (top) Standard incremental frequency and cumulative frequency plots. The GR extrapolation is given by the *b*-projection. (b) (middle) Straight lines represent three populations deduced from the composite (heavy curved line). Filled triangles represent the model of 1.5% of 4.1 ± 0.6, 21.3% of 1.9 ± 0.7, and 78.2% of 1.6 ± 0.3. (c) (bottom) The model given in *b*. (dashed line) is compared with the original incremental distribution (solid line with filled triangles). The recombination was calculated every 0.1 magnitude unit. The peak for the 4.1 population is less than 1.

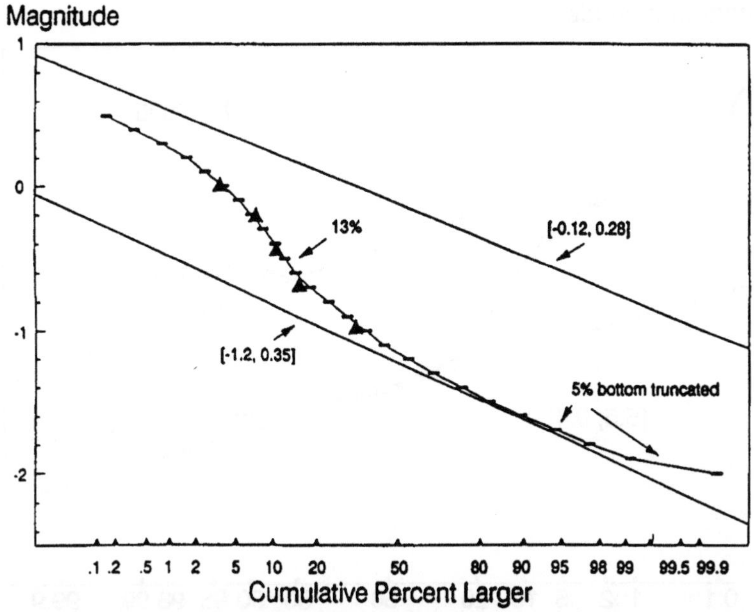

Figure 5. An example of multiple populations at the low magnitude end. The curved line is the aggregate probablity curve. The filled triangles indicate the fit by combining the two straight line populations as 13% of -0.12 ± 0.28 and 87% of -1.2 ± 0.35. The curvature upward at the low magnitude end is a pattern associated with bottom truncation of the data set. After Speidel and Mattson (1993b).

on the existence of data in the appropriate range. The GR relationship has no such limit once b is established.

At the small end of the magnitude spectra, the GR relationship does not fit most data sets. This is usually explained by incomplete data collection, but where data sets at small magnitudes are claimed to be complete, the GR relationship still does not hold, a situation expected on theoretical grounds (Lomnitz, 1974). For example, Trifu *et al.* (1993) gathered data in an underground mine and have clear enough control to document that the pattern is not due to instrumental insensitivity at low magnitudes. Two populations are deduced (Speidel and Mattson, 1993b) but no mechanisms were able to be assigned (Figure 5).

At the high end of the magnitude distribution, Figure 6 illustrates the probability graph for approximately 2600 earthquakes from the Pacheco–Sykes catalog for 1900–1982 (Pacheco and Sykes, 1992) supplemented by the 1983–1990 values from the Harvard CMT catalog. As first reported in 1992 (Mattson and Speidel, 1992) two populations of moment magnitude can be deduced from the catalog, 7.5 $\pm$ 0.5 and 5.5 $\pm$ 0.4. If the assumption that the different populations are related to differing mechanisms is correct (Speidel and Mattson, 1993b, 1995), the 7.5 $\pm$ 0.5 population should itself be a composite of large earthquakes with varying mechanisms of thrust, strike-slip, normal, and reverse. Isolating the population identified as having a strike-slip mechanism (Speidel and Mattson, 1993a), Figure 7

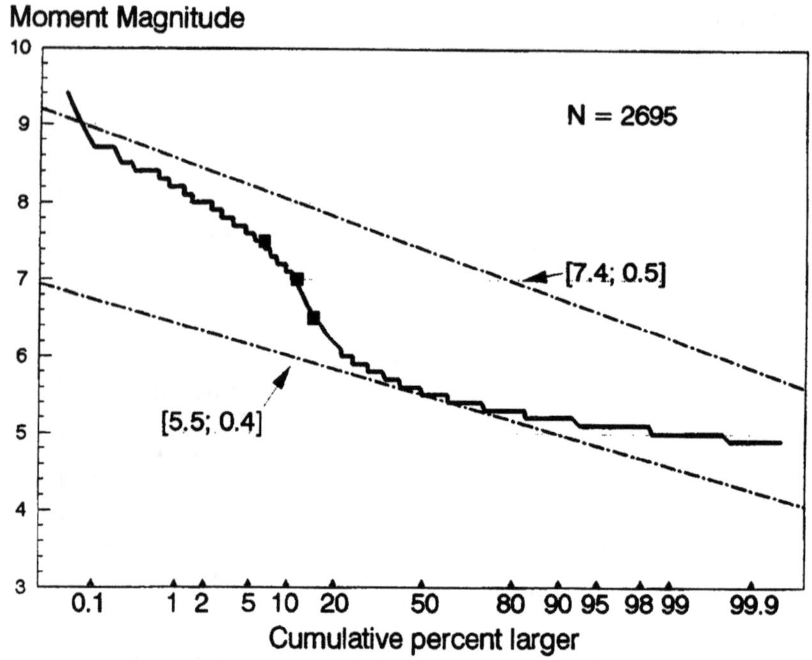

Figure 6. An example of two-population pattern at the high magnitude end. Filled squares indicate the fit by combining the straight line populations as 15% of 7.4 ± 0.5 and 85% of 5.5 ± 0.4. After Mattson and Speidel (1992).

illustrates that the 60 great (> M_w 7.0) strike slip earthquakes subset can indeed be described as a normal distribution. The population parameters are M_w = 7.2 ± 0.6 or, as a lognormal distribution of moment, $M_o = 0.74 \times 10^{20}$ Nm.

It thus appears that different mechanisms have distinct lognormal distributions. Region to region variation in either the density function and/or the cumulative distribution function could just be the effect of the occurrence of different faulting mechanisms or variation in proportion of activity of those mechanisms. Over 3000 earthquakes for 1965–1991 in Fennoscandanavia from what Ahjos and Uski (1992) term their 'instrumental data set' are plotted in Figure 8 (Speidel *et al.*, 1996). The three largest populations deduced, 3.45 ± 0.65, 2.3 ± 0.65, and 1.35 ± 0.5 with correlation coefficients less than 0.15, account for 98.5% of the total. The fit of the model to the data is indicated both in the density function plot (Figure 8a) and the cumulative distribution function (Figure 8b). The *b* value fit between M2 and M4 severely overestimates the occurrence of large magnitude earthquakes.

Figure 8c illustrate the frequency-magnitude relations for three regional subsets. Note that the straight line underestimates activity in one region while overestimating in two others. Figure 8d give the cumulative distribution function. The curves look quite different yet all can be described as linear combinations of the various deduced populations, albeit present in differing amount in each region. Figure 3, used earlier as an example of the pattern that can be interpreted as a mixture of

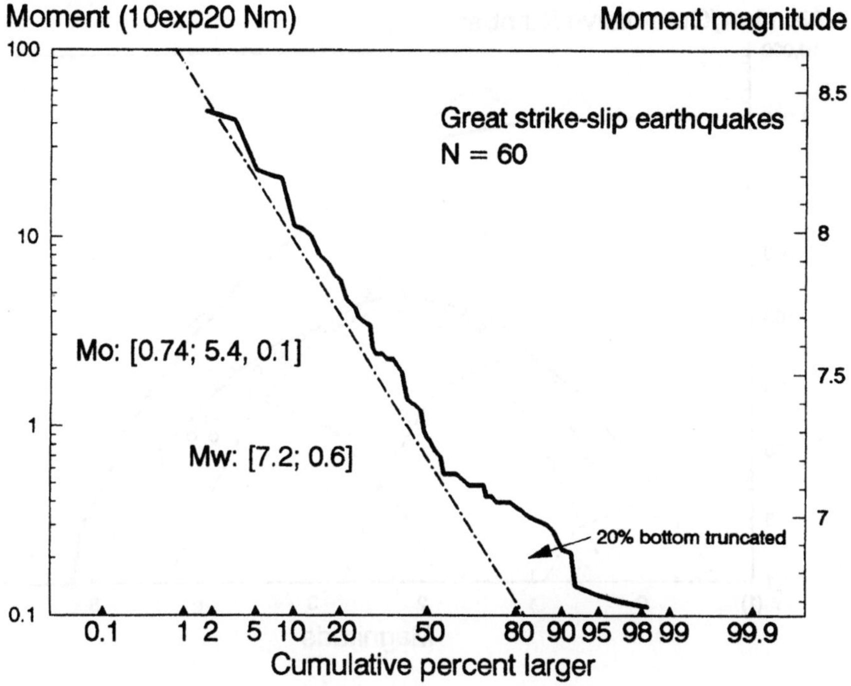

Figure 7. Pattern for Great Strike-slip earthquakes. Data from Pacheco and Sykes (1992). After Speidel and Mattson (1993a).

two normal distributions, is a subset of the Figure 8 data and a combination of I and II from Figures 8c and 8d. Information of this sort would be lost if only the G-R relationship were addressed. That is, total activity and mean magnitude are not sufficient to analyze frequency-magnitude composites.

3. Implications for PSHA

As mentioned in the introduction, PSHA can be carried out with any probability distribution, including those indicated in Figure 8b. To estimate the probability of occurrence for a particular magnitude, it is necessary not only to have the parameters of the populations, but also the activity in events per year for each population. This activity is estimated by dividing the number of earthquakes in the population by the number of years in the catalog. If populations overlap in magnitude (as can be seen they do in both Figures 8a and 8b), the contribution from each population has to be considered. Thus,

$$\lambda_M = \sum P_{i,M}\lambda_i \tag{2}$$

where λ_i is the activity of population i in number of earthquakes per year, $P_{i,M}$ is the probability of magnitude $\geq M$ for population i, and λ_M is the total activity for

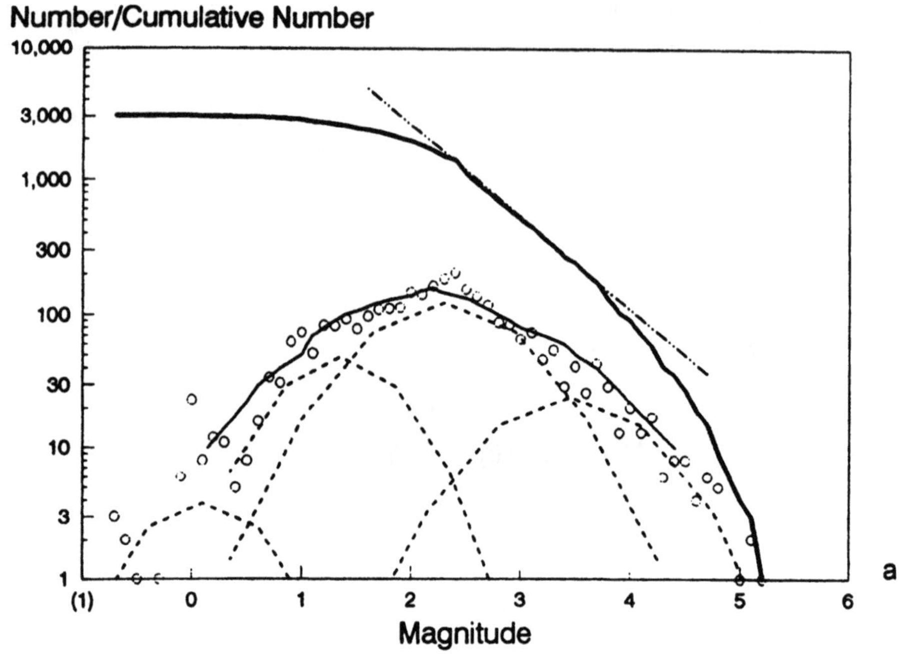

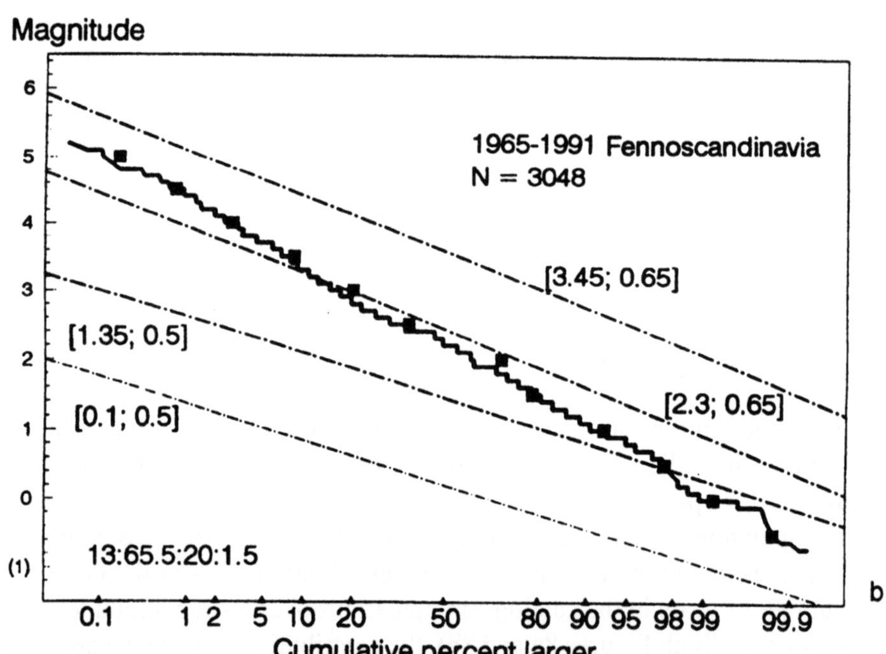

Figure 8a,b. Examples from the Fennoscandinavian catalog. Data from Ahjos and Uski (1992). After Speidel, Sy, and Mattson (1996). (a) Cumulative (solid triangles) and incremental (open circles) frequency plots for 3048 earthquakes. Deduced populations are given by dashed lines. Their sum by the solid line. One possible *b*-value is indicated by the dash-dot line. (b) Straight lines represent the populations deduced from the composite population given by the curved line of short dashes. Filled squares illustrate a model of 13% of 3.45 ± 0.65, 65.5% of 2.3 ± 0.65, and 20% of 1.35 ± 0.5.

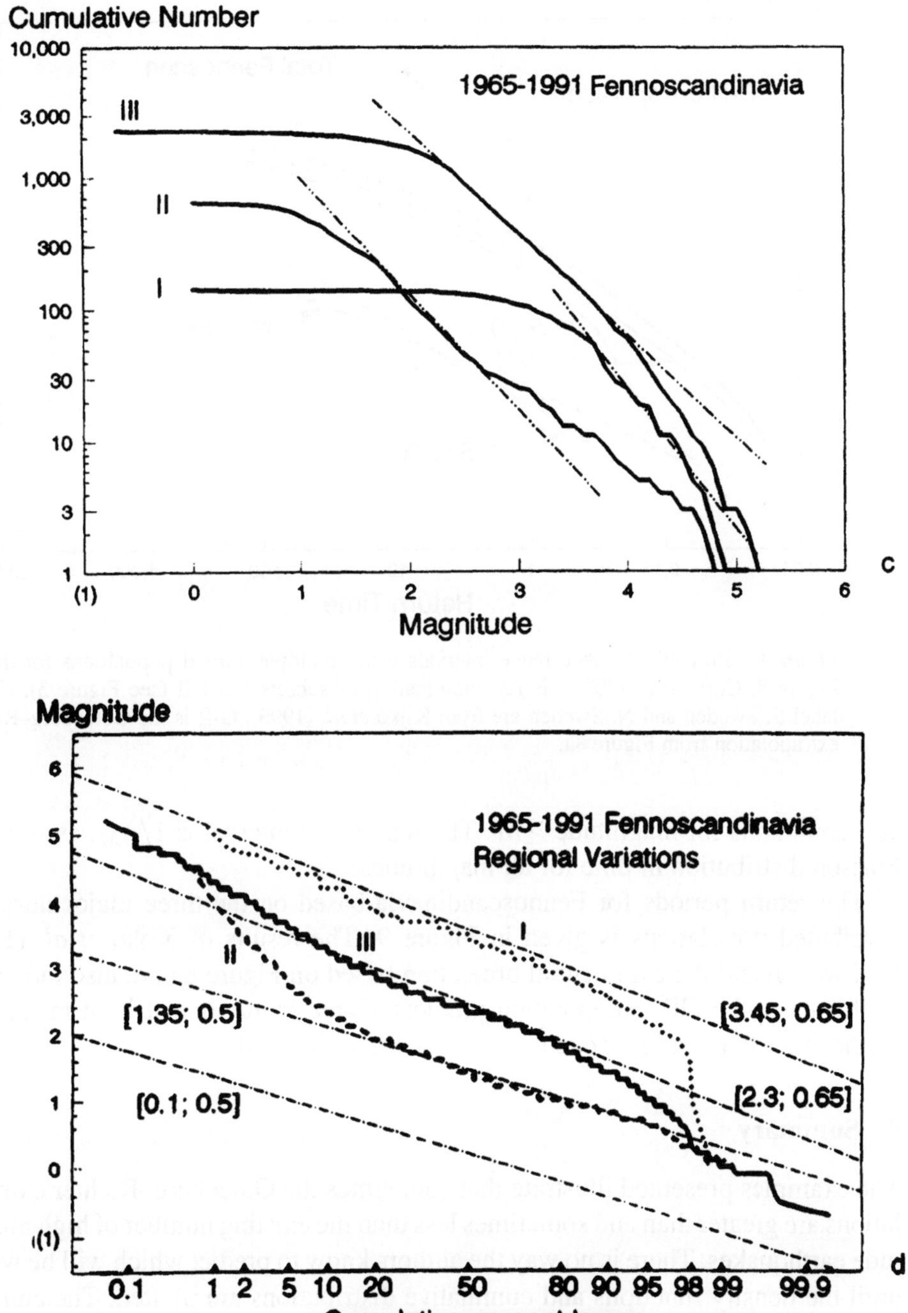

Figure 8c,d. Examples from the Fennoscandinavian catalog. Data from Ahjos and Uski (1992). After Speidel, Sy, and Mattson (1996). (c) Subsets of the data sorted by region. I = ≥70° N latitude and east of 0° longitude. II = ≥70° N and west of 0° longitude. III = <70° N. Straight dash-dot lines indicate slopes for *b*-values. (d) Cumulative probability plots for subsets I, II, and III. Straight dash-dot lines indicated the deduced populations.

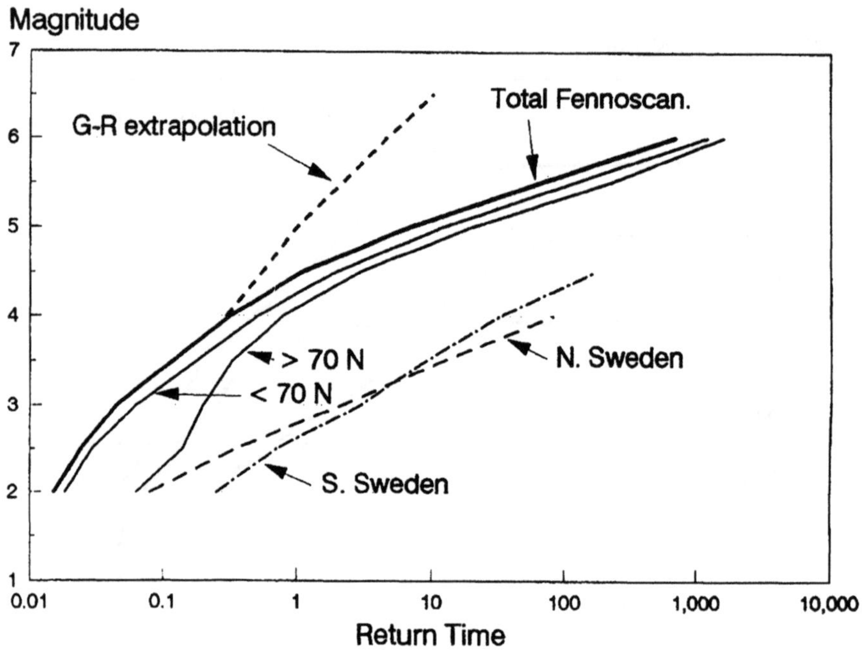

Figure 9. Probabilistic recurrence intervals using multiple normal populations for data of Figure 8. Curve for >70° N is for combination of subsets I and II (see Figure 3). Curves label S. Sweden and N. Sweden are from Kijko *et al.* (1993). G-R is the Gutenberg–Richter extrapolation from Figure 8a.

all populations for magnitude $\geq M$. The recurrence interval is $1/\lambda_M$, assuming a Poisson distribution in time for all magnitudes.

The return periods for Fennoscandinavia based on the three major normally distributed populations is given in Figure 9. The results of Kijko *et al.* (1993) for Sweden and the exponential projection based on Figure 8a are also indicated. In this case, the GR extrapolation predicts a greater number of high magnitude earthquakes than has ever occurred, as shown in Figure 8.

4. Summary

The examples presented illustrate that sometimes the Gutenberg–Richter extrapolations are greater than and sometimes less than the existing number of high magnitude earthquakes. There is no way the authors know to predict which will be which until the density functions and cumulative distributions are plotted. The cumulative distribution functions plots on normal probability graphs provide information about the gross structure of number and parameters of earthquake populations. This analysis, when coupled with fault mechanism where possible, gives a new view of what seismicity patterns might actually be.

The difficulty that the Gutenberg–Richter relationship shares with the Speidel–Mattson technique is that normal (or any other) distributions do not allow anything to be said about the probability of earthquakes whose magnitudes are not covered by the data (Speidel, 1996). That is, the problem with PSHA is that it depends on a frequency-magnitude model totally dependent on the pattern of pre-existing earthquakes. PSHA can not be justified on statistical grounds because the probability of an event that has not been observed can not be estimated. Extrapolating into a data region where no observations exist is scientifically unsound. The fact that laws and regulations may require such extrapolation adds a political requirement without resolving the scientific one. Quick analysis of the cumulative distribution function and determinations of the parameters of the high magnitude populations (or lack thereof) might provide a means for helping with the political part of the problem.

Acknowledgement

We thank the reviewers for their cogent arguments and suggested references.

References

Ahjos, T. and Uski, M.: 1992, Earthquakes in northern Europe in 1375–1989, *Tectonophysics* **207**, 1–23.

Aitchison, J., and Brown, J. A. C.: 1974, *The Lognormal Distribution*, Cambridge University Press, London.

Aki, K.: 1987, Magnitude-frequency relation for small earthquakes: A clue to the origin of f_{max} of large earthquakes, *J. Geophys. Res.* **92**, 1349–1355.

Båth, M.: 1981, Earthquake magnitude – Recent research and current trends, *Earth. Sci. Rev.* **17**, 315–398.

Bonilla, M., Mark, R., and Lienkaemper, J.: 1984, Statistical relations among earthquake magnitude, surface rupture, and surface fault displacement, *Bull. Seismol. Soc. Am.* **69**, 2003–2024.

El-Sayed, A. and Wahlström, R.: 1996, Distribution of the energy release, *b*-values and seismic hazard in Egypt, *Natural Hazards* **13**, 133–150.

Espinosa, S.: 1996, Probabilistic macroseismic hazard assessment for Nicaragua (preliminary results), *Natural Hazards* **13**, 179–202.

Frohlich, C.: 1989, The nature of deep-focus earthquakes, *Am. Rev. Earth Planet. Sci.* **17**, 227–254.

Frohlich, C. and Davis, S. D.: 1993, Teleseismic *b* values: or, much ado about 1.0, *J. Geophys. Res.* **98**, 631–644.

Giardini, D.: 1988, Frequency distribution and quantification of deep earthquakes, *J. Geophys. Res.* **93**, 2095–2105.

Green, H., II: 1993, The mechanism of deep earthquakes, *EOS, Trans. Am. Geophys. Union* **74**, 23.

Gutenberg, B. and Richter, C. F.: 1949, *Seismicity of the Earth and Associated Phenomena*, Princeton Univ. Press, Princeton.

Hough, S.: 1996, The case against huge earthquakes, *Seismol. Res. Lett.* **67**, 3–4.

Jackson, D.: 1996, The case for huge earthquakes, *Seismol. Res. Lett.* **67**, 3–5.

Kagan, Y.: 1996, Comment on 'The Gutenberg-Richter or Characteristic Earthquake, Which Is It?', by Steven G. Wesnousky, *Bull. Seismol. Soc. Am.* **86**, 274–285. Wesnousky response, 286–291.

Kijko, A., Skourdas, E., Wahlström, R., and Mäntyniemi, P.: 1993, Maximum likelihood estimation of seismic hazard for Sweden, *Natural Hazards* **7**, 41–57.

Krinitzsky, E.: 1993, Earthquake probability in engineering – Part 2: Earthquake recurrence and limitations of Gutenberg-Richter *b*-values for the engineering of critical structures, *Eng. Geol.* **36**, 1–52.

Krinitzsky, E.: 1995, *State-of-the-art for assessing earthquake hazards in the United States, Report 29, Selection of earthquake ground motions for engineering*, Misc Paper S-73-1, Waterways Experiment Station, US Army Corps of Engineers, Vicksburg, Miss.

Lomnitz, C.: 1974, *Global Tectonics and Earthquake Risk*, Elsevier, Amsterdam.
Lomnitz, C.: 1994, *Fundamentals of Earthquake Prediction*, Wiley, New York.
Main, I.: 1987, A characteristic earthquake model of the seismicity preceding the eruption of Mount St. Helens on 18 May 1980, *Phys. Earth Planet. Interiors* **49**, 283–293.
Mattson, P. and Speidel, D.: 1992, A model for nonlinear frequency-magnitude relationships for earthquakes, *EOS* **73**, Spring Meeting Suppl., 197.
Mattson, P. and Speidel, D.: 1994, Seismic hazard in the Central Mississippi Valley: the demise of *b*-values, (Abstract), *EOS* **75**(16), Spring Meeting Suppl., 236.
Mattson, P. and Speidel, D.: 1996, Improbable probabilistic seismic hazard analysis, (Abstract), Sixth International Symposium on Natural and Man-made Hazards, Toronto, Natural Hazards Society, *HAZARDS-96 Abstracts V.* **90**.
NRC (National Research Council), Panel on Seismic Hazard Analysis: 1988, *Probabilistic Seismic Hazard Analysis*, National Acad. Press, Washington, DC.
Nicol, A., Walsh, J., Watterson, J., and Gillespie, P.: 1996, Fault size distribution – are they really power-law?, *J. Structural Geol.* **18**, 191–197.
Pacheco, J., Scholz, C., and Sykes, L.: 1992, Changes in frequency-size relationships from small to large earthquakes, *Nature* **355**, 71–73.
Pacheco, J. and Sykes, L.: 1992, Seismic moment catalog of large shallow earthquakes, 1900 to 1989, *Bull. Seismol. Soc. Am.* **82**, 1306–1349.
Papadopoulos, G., Skafida, H., and Vassiliou, I.: 1993, Nonlinearity of the magnitude-frequency relation in the Hellenic arc-trench system and the characteristic earthquake model, *J. Geophys. Res.* **98**, 17,737–17,744.
Purcaru, G. and Zorilescu, D.: 1971, A magnitude-frequency relation for the lognormal distribution of earthquake magnitude, *Pure Appl. Geophys.* **87**, 43–54.
Ranalli, G.: 1975, A test of the lognormal distribution of earthquake magnitude, *Veroff. Zentralinst. Phys. Erde*, D.A.W. Potsdam, **31**, 163–180.
Reiter, L.: 1990, *Earthquake Hazard Analysis: Issues and Insights*, Columbia Univ. Press, New York.
Rydelic, P. and Sacks, I.: 1989, Testing the completeness of earthquake catalogues and the hypothesis of self-similarity, *Nature*, **337**, 251–253.
Sacuiu, I. and Zorilescu, D.: 1970, Statistical analysis of seismic data on earthquakes in the area of the Vrancea Focus, *Bull. Seismol. Soc. Am.* **60**, 1089–1099.
Scholz, C. H.: 1991, Earthquakes and faulting: self-organized critical phenomena with a characteristic dimension, in T. Riste and D. Sherrington (eds), *Spontaneous Formation of Space-Time Structures and Critically*, Kluwer, Netherlands, pp. 41–56.
Schwartz, D. and Coppersmith, K.: 1984, Fault behavior and characteristic earthquakes: Examples from the Wasatch and San Andreas fault zones, *J. Geophys. Res.* **89**(B7), 5681–5698.
Sinclair, A. J.: 1976, *Applications of Probability Graphs in Mineral Exploration*, Special Volume No. 4, Association of Exploration Geochemists, Rexdale, Ontario.
Sornette, D., Knopoff, L., Kagan, Y., and Vanneste, C.: 1996, Rank-ordering statistics of extreme events: Application to the distribution of large earthquakes, *J. Geophys. Res.* **101**, 13,883–13,893.
Speidel, D.: 1996, A hermeneutic reading of the seismic record: improbable probabilities?, *Eng. Geol.* **43**, 1–3.
Speidel, D. and Mattson, P.: 1993a, Magnitude-frequency distribution by mechanism for great earthquakes, *EOS* **74**, Fall Meeting Suppl., 436.
Speidel, D. and Mattson, P.: 1993b, The polymodal frequency-magnitude relationship of earthquakes, *Bull. Seismol. Soc. Am.* **83**, 1893–1901.
Speidel, D. and Mattson, P.: 1995, Questions on the validity and utility of *b*-values: An example from the Central Mississippi Valley, *Eng. Geol.* **40**, 9–27.
Speidel, D., Mattson, P., and Sy, B.: 1996, Seismic hazard analysis without the Gutenberg-Richter relationship, *Am. Soc. Civil Eng. Conference NDR '96*.
Speidel, D., Sy, B., and Mattson, P.: 1996, Seismicity patterns and earthquake recurrence in Scandinavia: Multiple populations, (Abstract), *Seismol. Res. Lett.* **67**, 55.
Taylor, D. W. A., Snoke, J. A., Sacks, I. S., and Takanami, T.: 1990, Nonlinear frequency-magnitude relationships for the Hokkaido Corner, Japan, *Bull. Seismol. Soc. Am.* **80**, 340–353.

Titterington, D., Smith, A., and Makov, U.: 1985, *Statistical Analysis of Finite Mixture Distributions*, Wiley, Chichester.
Trifu, C., Urbancic, T., and Young, R.: 1993, Non-similar frequency-magnitude distribution for $M < 1$ seismicity, *Geophys. Res. Lett.* **20**, 427–430.
Wells, D. and Coppersmith, K.: 1994, New empirical relationships among magnitude, rupture length, rupture area, and surface displacement, *Bull. Seismol. Soc. Am.* **84**, 974–1002.
Wesnousky, S.: 1994, The Gutenberg-Richter or Characteristic Earthquake Distribution, Which is It?, *Bull. Seismol. Soc. Am.* **84**, 1940–1959.
Zipf, G.: 1949, *Human Behavior and the Principle of Least-Effort*, Addison-Wesley, Reading, Mass.

Natural Hazards **16:** 181–202, 1997.

Atmospheric Hazards Preparedness in Bangladesh: A Study of Warning, Adjustments and Recovery from the April 1991 Cyclone

C. EMDAD HAQUE
Department of Geography, Brandon University, Brandon, Manitoba, Canada R7A 6A9

(Received: 5 December 1996; in final form: 6 November 1997)

Abstract. In probabilistic terms, Bangladesh is prone, to at least one major 'tropical cyclone' every year. This situation is primarily due to the geographical location of Bangladesh in tropical Asia, and to its concave coastline and shallow continental shelf. The devastating impact of such cyclones on humans stems from a combination of intense human occupation of the area, predominance of traditional sociocultural values and religion, the precarious socioeconomic conditions of the majority of the coastal inhabitants, and the lack of a coordinated institutional disaster planning and management strategy. Bangladesh has experienced several catastrophic environmental disasters during the last decade; among these events, the 1991 April cyclone was the most catastrophic in terms of both physical and human dimensions.

An initial study was carried out in the coastal regions of Bangladesh less than two weeks after they were hit by the severe cyclone of 29 April 1991. This research examined the process through which warning of the impending disastrous cyclone was received by the local communities and disseminated throughout the coastal regions of Bangladesh. It was found that the identification of the threatening condition due to atmospheric disturbance, the monitoring of the hazard event, and the dissemination of the cyclone warning were each very successful. The present study followed up on the initial research by surveying 267 respondents with an elaborate survey instrument, focusing on the most crucial academic and planning issues identified in the 1991 study. In particular, the nature and characteristics of the cyclone preparedness of the coastal inhabitants were assessed by the study; other factors considered included rural-urban variations, mainland-island differences, the nature and role of previous knowledge, and the disaster experience.

The survey results show the variety of indigenous adjustment mechanisms that help to rehabilitate the survivors; also visible are the profound roles played by the social inequality variables and the magnitude of physical vulnerability in influencing the disaster loss and recovery process. The study recommends that hazard mitigation policies should be integrated with national economic development plans and programs. Specifically, it is suggested that the cyclone warning system should incorporate the human response to warnings as its constituent part, and in this way accommodating human dimensions in its operational design.

Key words: cyclone, forecasting, warning, adjustment measures, response, recovery, relief and rehabilitation, Bangladesh, urban/mainland, offshore island.

1. Introduction

The geography of Bangladesh, as a humid tropical region, has largely caused the country to be extremely susceptible to catastrophic disasters such as floods, cyclones and storm surges; these environmental phenomena complicate the already precarious socioeconomic and demographic conditions of the country. In probabilistic terms, Bangladesh is prone to at least one major tropical cyclone every year (Mooley, 1980), but an analysis for the period of 1890–1969 shows that, on average, 13 depressions were generated annually, of which 4.6 transformed into tropical cyclones (Raghavendra, 1973). A number of meteorological conditions that are necessary for the formation of tropical cyclones exist in the Bay of Bengal. In the bay, the water temperature remains at least 26°C throughout the year (Hastenrath and Lamb, 1979); this factor is probably of greatest importance as sea temperature of this magnitude to a depth of about 60 m is needed in the area for the development of a cyclone (Anthes, 1982; Emanuel, 1988). The second general requirement is a minimal vertical shear in the horizontal wind near the prospective cyclone. This situation exists in the Bay of Bengal during two periods of the year, namely, April–May and September–November, when the wind field over the bay is conducive to the necessary low-level inflow and high-level outflow. In addition, the shallowness of the deltaic coastal shelf contributes to surge elevation (Johns *et al.*, 1983), while the concavity of the bay and its estuaries further amplifies the surges (Dube *et al.*, 1982).

Encompassing an area of 147 570 km^2, Bangladesh presently contains a population of more than 118 million. The per capita income is less than $220 U.S. per annum and the majority of the rural inhabitants are functionally landless. Consequently, neither the institutions (i.e., collective entities) nor individuals can afford to implement capital-intensive preventive measures to avoid or minimize hazard-loss. Modifications in preparedness may therefore significantly improve the nation's ability to mitigate disasters.

Hazard warning systems (Sims and Baumann, 1983) and preparedness programs have generally proven to be effective in the developed world for informing people of impending danger as well as for modifying resource use systems (Smith, 1992: 198–199). However, since the reactions of people to hazard warnings have not been adequately and systematically investigated, literature on the overall effectiveness of the warning systems is meager. This study is primarily intended to examine the different components of the existing warning system in Bangladesh, the individual and institutional preparedness and adjustment measures to cope with cyclones, and the nature and characteristics of the disaster-recovery process.

2. Hazard Warning and Mitigation Measures: Conceptual Considerations

In general, reduction of losses due to extreme physical events can be achieved in two ways. The first option is the preventive 'environmental control', which involves modifying the hazard event itself. However, this type of large-scale environmental control is viable in only a few cases; usually, the suppression of 'natural events' is either impossible or, at best, entails uncertain side effects (Smith, 1992: 81). Also, due to economic, social, political and technological constraints, an absolute security against natural hazards can hardly be realized. For this reason, the modification of human-made structures to make them hazard-resistant tends to be confined to small-scale interventions, such as developing and implementing building codes and other local regulations.

The second method of reducing loss involves changing human attitudes and behaviors towards nature. This approach, termed as 'human vulnerability modification', integrates technological means and human elements for hazard loss reduction. Smith (1992) has attempted to group the range of vulnerability modification options into the following categories: (i) forecasting and warning; (ii) community preparedness; and (iii) land-use planning. Due to the complete absence of an organizational structure to regulate and implement land-use planning in rural Bangladesh, home to the majority of the population, the scope of the third option is extremely limited in this country. The first two categories are then the focus of the present study because of their relevance to the context of Bangladesh.

The main purpose of warning systems is two fold: to enable people to take precautionary measures to minimize loss, and to insist on evacuation of an area in advance of an approaching hazard; success depends on the reactions and responses of people. Human responses to these warnings involve a complex set of sociopsychological, socioeconomic and cultural variables, which together lead to a wide variation in reactions, even including a reluctance to either take precautionary action or evacuate. In this connection, psychological block factors that might hamper effective warning responses are stressed by some investigators (Williams, 1964). In the face of this complexity, an assessment of responses of those being warned is crucial to develop an effective feedback system, considering that delays or halts in the flow of messages have been cited as problems both in the developed and the developing world (Warrick *et al.*, 1981). Consider, for example, the problematic cyclone warning which was issued from the Meteorological Center at Gandhinagar, India to alert the population along the coastal districts of Gujarat in December of 1981. Sinha and Avrani (1984) found that a lack of coordination among the three responsible authorities (i.e., the civil administration, the port authority, and the observatory) left hundreds of thousands of fishermen and settlers uninformed and thus at the mercy of nature.

Although response to a hazard is largely related to perception of the threat and to awareness of opportunities to make adjustments, like the response to warnings, it is also profoundly influenced by socioeconomic and cultural constraints (Haque and Zaman, 1990; Haque, 1988; 1991). Human response and adjustment research in Bangladesh has focused primarily upon floods (Rasid and Paul, 1989; Haque and Zaman, 1993), riverbank erosion (Haque and Zaman, 1990; Mamun, 1996), droughts (Paul, 1992; Rahman, 1995), and storm surges (Islam, 1971; 1974; 1992; Khalil, 1992). By its absence from this list the nature of response to high winds and cyclones in the country has remained relatively unknown. Also, the degree to which individual resource managers and users are active in undertaking pre-disaster adjustment measures, as well as how they react during the onset of the event, have remained unexplored in Bangladesh disaster studies.

Other gaps are noticeable in relief efforts after a disaster. The issues concerning long-term effects due to disaster occurrences, recovery and rehabilitation of the victims at the local level usually do not receive continued attention from the concerned agencies. One of the reasons for such a pattern appears to be that the investment in disaster rehabilitation and recovery pays-off mainly in the long-term. Another significant reason is that, in most cases, recovery and rehabilitation measures, as undertaken by the development, emergency, relief and rehabilitation agencies and government ministries, are intended to address the welfare of the people. Such measures, in turn, deal with the problem of social opportunity costs involved with disruptions in normal functions of life and economic activities, rather than direct economic costs. Overall, both methodological difficulties and the lack of due institutional interest in the long-term recovery and rehabilitation process have left this a relatively neglected aspect in natural hazards research.

Palm (1990) has assessed the nature and contributions of hazards research by different disciplines of the social sciences. She criticizes the functionalist approach for its narrow scope; found mainly in sociology, it places its sole emphasis on organizational and community-related issues during the onset and post-event period. In contrast to this, the array of hazards studies in geography, anthropology and ecology in the past two decades indicates that a broad ecological framework is more useful to explain the complexity of environment-society relationships. In predictive terms, this latter approach can provide a more effective analytical tool by encompassing all phases of disaster process. The ecological framework prescribes the study of the characteristics of individual, group and institutional preparedness, and their responses and resilience to hazards. Specific disasters in Bangladesh will now be examined by using this framework.

3. Cyclone Preparedness in Bangladesh: The National Warning System

3.1. THE WARNING SYSTEM

As noted earlier, the Bay of Bengal which forms Bangladesh's coastline is one of the world's most active areas for the development of tropical low pressure systems. The cyclone of November 1970 has been the worst of this century, causing the deaths of 225 000 people, loss of 280 000 cattle, and destruction of properties worth U.S. $63 million (Frank and Hussain, 1971: 439). Development of this cyclone was initially noticed by the meteorological service of India and later detected by satellite and radar at Cox's Bazaar (Burton *et al.*, 1993), but no warnings could reach the rural communities in coastal districts.

Following the catastrophic cyclone, a plan was formulated to protect against similar kinds of events which may occur in the future. The Bangladesh Meteorological Department (BMD) presently now takes responsibility for preparing all weather forecasts and disaster warnings. These tasks are accomplished through five interconnected subdivisions, all coordinated by the Storm Warning Center (SWC): (i) observations, (ii) communications, (iii) display and manipulations, (iv) analysis,

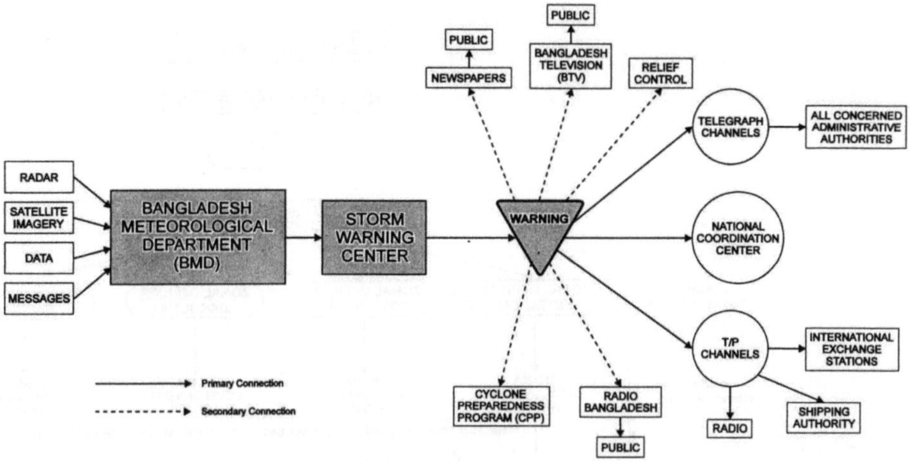

Figure 1. The flow of atmospheric information administered by the Bangladesh Meteorological Department.

and (v) prognosis (i.e. preparation of forecast) (Figure 1). The Center gathers data from four different sources. There are 35 surface observatories in the country which collect hourly data and send them to the Center in Dhaka; these data include wind speed, direction, humidity, air temperature and other meteorological variables at different elevations. The BMD also has three radar stations, at Dhaka, Khepupara and Cox's Bazar, transmitting hourly and half-hourly data (Figure 3). The Center receives U.S. NOAA (National Oceanic and Atmospheric Administration) satellite imagery via an earth station in Chittagong, and data from NOAA satellites 10 and 11 and the Japanese satellite GMS-4 via the Bangladesh Space Research and Remote Sensing Organization (SPARRSO). The World Meteorological Organization network also provides data to the BMD. The current organizational structure to procure, process, and disseminate atmospheric information, as shown in Figure 1, is efficient and effective. The nature and degree of coordination and cooperation between different subsets of the structure have improved due to regular review and feedback.

When the Storm Warning Center determines an impending threat, storm and cyclone warnings are disseminated to three areas: seaports, river ports, and the public. The Center also sends warnings directly to the National Coordination Committee (NCC), chaired by the Prime Minister, with representatives from the Cyclone Preparedness Program (CPP), user agencies (such as the Bangladesh Red Crescent Society, health administration, relief and rehabilitation authorities, and non-governmental organizations), mass media, and local administration.

The Cyclone Preparedness Program, an important link in the national warning system, was established in 1972 under an agreement between the Bangladesh Red Crescent Society (BRCS) and the Government of Bangladesh; it aims to undertake effective cyclone preparedness measures in the coastal belt prevent the loss of more

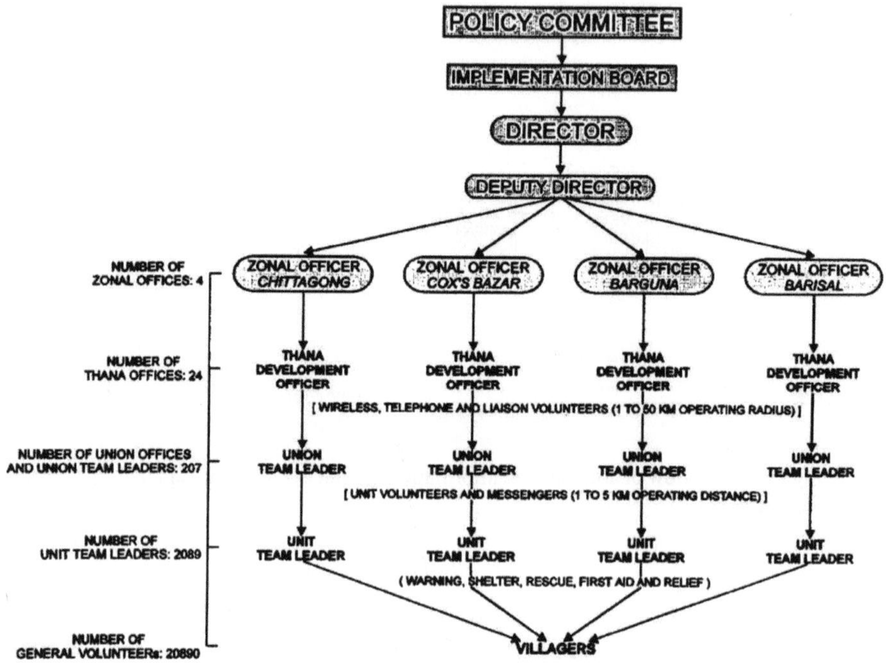

Figure 2. The organizational structure of the CPP.

than 70 million lives. The administrative structure of the CPP is depicted in Figure 2. Presently, the CPP has 20 890 trained volunteers in 2,089 units, divided among 207 unions under 24 *thanas* (subdistricts) of eight coastal districts, which further divide into four zones. Each union roughly has 10 units, individually involving at least 10 volunteers – usually these include school teachers, social workers, *moulavis* (clergymen), local government officials, and community leaders (Figure 2). At the zonal offices volunteers receive daily CPP communications, and in the case of potential hazards, disseminate warning signals, assist people in the evacuation process, execute rescue operations, provide first aid and help in the distribution of relief goods.

As a further measure, the national government of Bangladesh provides specific directives in the Standing Orders for Cyclones to ensure speedy and systematic management of any emergency situation. Several administrative groups are given specific functions and responsibilities during pre-disaster, disaster occurrence and post-disaster stages; these include: the concerned ministries, regional administrations (such as divisional commissioners), district commissioners, local administrations such as (*thana parishads* or councils, union *parishad*) and the Bangladesh Army, Navy, and Air Force. The National Coordination Committee (NCC) in Dhaka regulates the necessary steps as the central national body, while the Ministry of

Relief and Rehabilitation coordinates the post-disaster activities in collaboration with the BDRCS and other voluntary and international agencies.

3.2. THE 1991 CYCLONE WARNING PROCESS

On 24 April 1991, the BMD first recognized a low pressure system as a potential meteorological hazard at 2400 Greenwich Mean Time (GMT) (i.e., 06:00 a.m. local time) the system was then located near 10° N, 90° E, with sustained winds of 55 kph and gusts to 75 kph (Climate Analysis Center, 1991; Figure 3). At 1200 GMT (i.e., 06:00 p.m. local time) on April 27 the system was upgraded to a tropical cyclone, with winds of 130 kph gusting to 160 kph. This message was transmitted to CPP field officers, and the volunteers were immediately moved to alert the public with house-to-house contact . When the storm curved towards the northeast at 0600 GMT (i.e., 12:00 a.m. local time) on 28 April the SWC advised hoisting of 'local cautionary signal' number 3. Altogether, there are eleven signals, the most extreme indicating that the warning and communication system may completely collapse due to the impending disaster. Signal number 3 implies wind gust in the range of 40 to 50 kph. At 0800 GMT (i.e., 02:00 p.m. local time) on the same day the SWC upgraded the warning to 'local warning signal' number 4, and this message was transmitted to the CPP field officers. This signal indicated that wind would gust between 51 and 61 kph and that ports could be affected by the event (Mirza and Pal, 1992). Field officers were further advised to organize meetings of the coordination committees at all administrative levels and to take all possible precautionary measures in the face of the impending cyclone.

As the situation worsened, the mass media joined the effort to alert the public. Radio Bangladesh and Bangladesh Television broadcasted warnings from the morning of 27 April. On 29 April, the media cancelled all of their normal schedules and concentrated on issues of cyclone preparedness, including evacuation procedures, disaster recovery, first aid and emergency measures. The daily newspapers, however, did not print warnings as the headline events, although the immediacy and serious nature of the threat was indicated in the text of the newspaper coverage.

The cyclone reached its maximum strength on 29 April, with sustained winds estimated at 260 kph, gusting to 315 kph. At 0400 GMT (i.e., 10:00 a.m. local time) the SWC advised 'Great Danger Signals' to all communication channels. The CPP field officers hoisted three signal flags in all the alert points and suggested that people move to safer places; before the onset of the catastrophic event the local volunteers evacuated more than 350 000 people to 308 cyclone shelters.

At around 1800 GMT (i.e., 12:00 p.m. local time) on 29 April, the cyclone made landfall along the Cox's Bazar and Chittagong coastline. This depression of pressure, in conjunction with the full moon of 28 and 29 April, resulted in the highest level of the normal range of pre-surge tides (Haque and Blair, 1992). The central pressure of the 1991 cyclone was at least as low as 938 mb, according to measurements taken from ships at anchor outside the port of Chittagong (Seven-

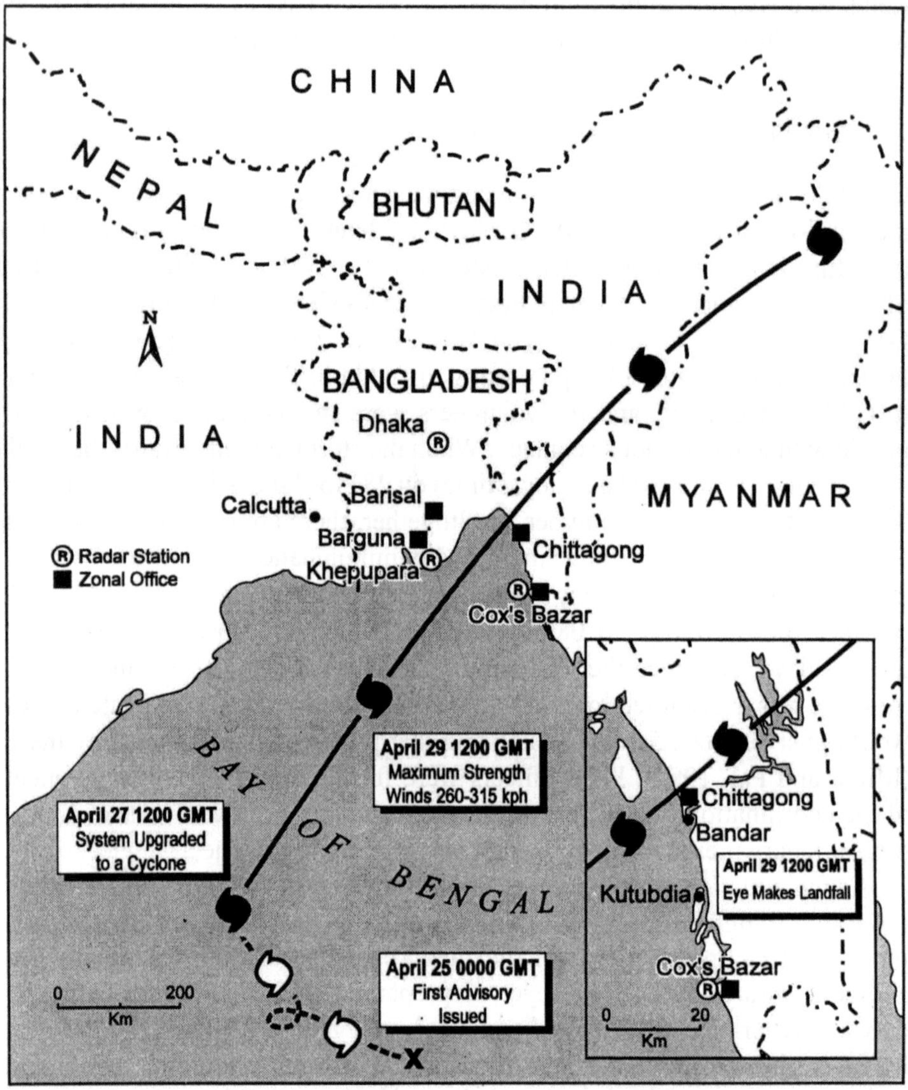

Figure 3. Location and path of the April 1991 cyclone into coastal Bangladesh.

huysen, 1991). Consequently, the surge and its related waves exceeded a height of six meters in the offshore islands and submerged more than 160 km of the country's coastline, including vast areas of Cox's Bazar, Chittagong, Bhola, and Noakhali districts.

The official death toll was recorded at 138 868, of which 23 were CPP volunteers who died while performing their duties. The government also recorded more than 460,000 injuries; among them 775 were CPP volunteers. The unprotected islands lost 40 to 50 percent of their population, while islands protected by embankments suffered a 30 to 40 percent loss of life – this in contrast to affected mainland

Table I. List of damages caused by the 1991 cyclone

Total affected districts	19
Total affected subdistricts (*thanas*)	102
Total affected municipalities	9
Total number of affected population	10 798 275
Damages to crops (in hectares)	
(i) fully	53 975
(ii) partly	320 607
Total number of houses damaged	
(i) fully	819 608
(ii) partly	882 750
Total number of deaths	138 882
Total number of people injured	139 054
Number of people missing	1225
Total number of livestock killed	
(cows, goat, sheep, poultry included)	1 061 029
Total number of educational institutions damaged	
(i) fully	3865
(ii) partly	5801
Total earthen-roads damaged (in kms)	
(i) fully	1230
Total number of bridges, culverts damaged	
(i) fully	496
Total number of embankments damaged (in kms)	
(i) fully	196
(ii) partly	941

Source: Islam *et al.*, 1992.

areas, which suffered a 20 to 30 percent fatality rate (Sevenhuysen, 1991). The estimated economic loss, according to the National Coordination Committee, was more than two billion dollars U.S. Agricultural losses were recorded at 1 061 029 head of cattle, along with the complete destruction of 51 108 hectares of crops and partial damage of another 156 500 hectares (Crossette, 1991; Sevenhuysen, 1991; BDRCS, 1992). Table I presents a list of damages caused by the devastating cyclone.

4. Community Preparedness and Individual Responses: Sample Survey Results

In order to gain insight into the varying responses to the 1991 cyclone, two distinct communities were surveyed in the months of June and July of 1992, one from an urban/mainland area, and one from a rural/offshore island area. It was presumed that variations in communication and infrastructural facilities might result in cyclone warnings being received more readily in the urban communities than in the offshore and rural communities. A similar survey was carried out by the author just after the April 1991 cyclone (Haque and Blair, 1992), but the sample size was limited to only 117 due to the unsuitable psychological state of the cyclone victims. The 1992 study was carried out with a larger and more authentic sample design in order to substantiate the validity of some of the previous findings.

In the 1992 survey, household units were considered as primary sampling units (PSU). Only the heads of the households were directly interviewed with a questionnaire. The PSUs were then drawn randomly to ensure the best possible representation of the target population. A total of 268 PSUs were covered – 162 from the urban/mainland sample and 106 from the offshore/rural sample. The urban/mainland sample was taken from the Bandar *thana*, consisting of the southwestern part of the City of Chittagong (Figure 3). The area felt severe effects, particularly by staggering economic losses, since it contains the major sea port facilities of Bangladesh, along with major industrial and commercial complexes. The rural/offshore island sample was obtained from the Kutubdia Island (Figure 3), a *thana* administrative area located 90 km south of Chittagong, along the northwestern coastline of Cox's Bazar district.

4.1. WARNING RECEPTION AT THE LOCAL LEVEL

The public received the hazard warning quite efficiently prior to the April 1991 Cyclone. The Chittagong municipal authorities, the district administration and nongovernmental organizations were first informed about the impending cyclone on 28 April, both by the NCC and the local CPP. Easy access by both paved and unpaved roads facilitated a rapid mobilization of the BDRCS volunteers, who reached all local 'wards' and communities of the Bandar *thana* area. Some relatively distant communities, settled mostly by poor fishermen, were not easily accessible, but the use of megaphones allowed these people to hear the warnings. The local mobilization of volunteers in Kutubdia began on 29 April. Although the Kutubdia *thana* headquarters is quite isolated from the mainland information sources, it does have a wireless center, operated by the CPP, which receives messages primarily from the surrounding *thana* CPP centers, or the regional centers stationed in Cox's Bazar. All of the six unions of the *thana* area were covered in warning dissemination efforts, including the villages in remote unions. The most distant communities experienced delays, especially those in Dhurong and North Dhurong, where they received cyclone warnings as late as the afternoon of April 29.

Among the survey respondents, more than 95 percent were present in the survey areas during the April 1991 cyclone. Out of a total of 162 respondents in the Bandar sample, only seven were out of the area; similarly, only six, out of a total of 106, were not on the Kutubdia island during the catastrophic event. Again, the communication system worked well all but five of the surveyed heads of households in Bandar received an early warning about the cyclone. Of all those who

Table II. Primary sources of early warnings of the 1991 cyclone

Source of early warnings	Bandar (urban/mainland)		Kutubdia (rural/offshore)	
	N	%	N	%
Radio broadcast	98	65.3	65	65.0
BDRCS	19	12.7	22	22.0
Newspapers	nil	–	nil	–
Local government officials/ community leaders	13	8.6	3	3.0
Neighbours	10	6.7	8	8.0
Other sources	10	6.7	2	2.0
Total	150	100.0	100	100.00

Source: Author's field survey, 1992.

heard the warning in both urban/mainland and rural/offshore island communities, about two-thirds received their warnings from radio broadcasts. Other aspects of the system were more effective in particular areas. The announcements and personal communications by the BDRCS volunteers were a more important source of information in the offshore islands than in the mainland, while local government officials and community leaders were more important in urban areas (Table II). It is observed that the warnings were widely received in both communities, but from a limited number of primary sources. Overall, the warning dissemination system that had been active in coastal Bangladesh since 1972 could be credited as very efficient. This success compares favorably to the enormous management problems of warning dissemination in India; as stated earlier, Sinha and Avarni (1984) cited a failure in passing out an early warning during the 1981 Gujarat cyclone in India.

4.2. RESPONSE TO WARNING AND ADJUSTMENT MEASURES

Responses to warnings among the coastal inhabitants were quite varied. In some instances, all the members of a household evacuated their usual place of residence if they were not residing in masonry construction. In some other cases, only the most vulnerable members such as women and children were evacuated. The remaining population stayed at home, and opting either for some alternative emergency measures or inaction, thereby accepting the potential loss and damage. The survey results indicate that about 49 percent of the urban/mainland and 71 percent of the rural/offshore island sample took this route, avoiding any deliberate emergency action after receiving the cyclone warning (Table III). Thus, the vast majority of the potential hazard victims remained vulnerable to the cyclone and storm surge strike.

A combination of factors contributed to such *en masse* inaction in the face of a severe threat, however, three of explanations stand out as the principal factors.

Table III. Distribution of measures taken by respondents after receiving early warning and before onset of flooding (multiple response possible)

Hazard mitigation measures taken	Bandar (urban/mainland) $N = 162$		Kutubdia (rural/offshore) $N = 106$	
	Frequency	%	Frequency	%
A. *'Active' measures*				
All household members evacuated	23	14.2	11	10.4
Women and children evacuated	22	13.6	5	3.1
Household goods and utensils moved	13	8.0	5	4.7
Household goods stored underground for future recovery	4	2.4	5	4.7
Anchorages placed	4	2.4	1	0.9
Livestock set free	3	1.9	18	16.9
Livestock moved to safer places	3	1.9	1	0.9
B. *'Passive' measures*				
No measures taken and prayed to to 'Allah'	79	48.9	75	70.7

Source: Author's field survey, 1992.

These are: (1) fatalism; (2) disbelief in the warnings; and (3) fear of loosing household assets (i.e., fear that home would be looted if abandoned for cyclone shelters) (Table IV). The unqualified acceptance of 'inevitable as inevitable' (Ittelson *et al.*, 1974) was, and generally has been, one of the common factors for staying home and not responding to a warning. Disbelief of warnings, due to experience of false warnings in numerous past occasions, was another principal factor causing inaction against warnings. The surveyed communities had received alarms frequently over the last two decades, particularly after the establishment of the early warning system in 1972. Due to the random and unpredictable nature of the track of tropical cyclones, as well as uncertainty associated with the life cycle of these meteorological disturbances, early warnings in many cases had appeared to be 'false' to the local communities. Since people gradually become immune to alarms, it may be suggested that there is a trade-off between the gains of advance warning dissemination even when the probability of hazard is low, and the gains of reducing false alarms to enhance response to high-probability warnings. As discussed earlier, the storm signal numbers issued by the CPP indicate the degree of threat and the required precautionary measures. However, to what extent such distinctions are transmitted to the local units is not fully known. Further research is required for a better understanding of this issue.

In addition, a low frequency of events may discourage people from undertaking precautionary measures, and thus leave them unprepared for coping with hazards.

Table IV. Reasons for staying home by household heads after receiving cyclone warning (multiple response possible)

Reasons stated	Bandar (urban/mainland) $N = 139$		Kutubdia (rural/offshore) $N = 95$	
	Frequency	%	Frequency	%
Did not believe warning	69	49.6	34	35.8
Fear of burglary	74	53.2	56	58.9
Thought that 'it would not happen here'	24	17.3	13	13.7
Thought house would provide protection	21	15.1	9	9.5
Too many people at shelters, embankments, and high grounds	10	7.2	18	18.9
Cyclone was 'Allah's will'	87	62.6	86	90.5

Source: Author's field survey, 1992.

In the case of Chittagong and Cox's Bazar districts, many young adults had not been exposed to any severe tropical cyclones before 1991, and only the elderly had experienced the last catastrophic cyclone of 1960. A common statement of the survey respondents was, 'It never happened here – we thought it's not going to happen this time, too'.

Almost half of the urban sample and slightly over one-third of the rural sample stated that people stayed home because they were afraid of having their homes looted during the evacuation (Table IV). This response implies that a denser network of smaller shelters would be preferable to less numerous larger shelters, since such a policy would reduce the house-to-shelter distance, allowing for better protection of home property. There was also an inadequacy of cyclone shelters, particularly in the offshore islands. In Kutubdia, about one-third of the respondents indicated that the lack of shelter did not allow them to evacuate females and children.

Aside from the evacuation option, a few other alternative emergency measures have been used to minimize hazard-loss. Setting livestock free to provide them an opportunity for survival was a preferred action among the offshore inhabitants while saving household goods received priority among the urban dwellers (see Table III). It has been observed that potential hazard victims tend to take emergency measures primarily under the threat of intensive stress (Hewitt and Burton, 1971), which implies that the opportunity cost of increased risk associated with delaying is measured in terms of post-event predicament.

As noted in Section 3.2, the Bandar *thana* area was devastated by the vigor of the wind and the accompanying storm surge. Many residential buildings were severely damaged, and roofs of factories, commonly made of corrugated iron sheets were blown away (Figure 4). Furthermore, communication and electric poles were broken, most trees in the area were uprooted, and hundreds of boats and dinghies,

Figure 4. A retail shopping area in Bandar *thana* where cyclonic winds destroyed roofs and walls. Masonry buildings often cannot withstand the high-velocity winds. Although material losses due to cyclones are high in urban areas, the loss of human lives is comparatively lower here than in rural and island communities.

locally known as *sumpans*, were completely demolished (Figure 5). Consequently, the magnitude of damage and economic loss was immense, affecting almost all industrial and commercial establishments. However, the loss of human lives was substantially lower in these urban communities, compared to the rural and isolated islands. Overall, the urban areas felt the benefits of better preparedness among organizations and individual household owners, superior infrastructure, and quicker response by the potential victims.

The storm surge was much more intense in the islands since these communities are more exposed to the Bay of Bengal and surrounded by shallow shorelines. High gusts of wind, indeed, caused severe economic and material damage, but the loss of human lives was, by and large, the result of people being literally washed away by the tide of the storm surges. As a result of these factors, 58.5 percent of the Bandar sample lost at least one family member due to the impact of the cyclone and its associated storm surges. By contrast, only 8.6 percent of the Bandar sample experienced a similar loss. In both locations, more than 60 percent of such human casualties were women. Furthermore, in Kutubdia the most significant damage was done to housing structures which are normally made of a combination of clay and earth, bamboo, thatch, and corrugated iron sheets (Figure 6). Other economic losses

Figure 5. A shipwreck in the outport near Chittagong. Although the amount of economic value is considerable, such losses are generally recoverable from insurance.

included livestock, poultry, trees, crops, household goods, farming equipment, boats, fishing nets and other equipment. Economic losses at the household level in Kutubdia also were much higher than in Bandar, which was also reflected in the disaggregated data by economic-loss groups (Haque, 1993).

4.3. POST-DISASTER RESPONSES: RELIEF AND REHABILITATION

With regard to the resilience and rehabilitation of disaster victims, linkages between the institutional and individual spheres are crucial (Palm, 1990; Burton *et al.*, 1993; Hewitt, 1983). The analysis of data on emergency relief and long-term rehabilitation measures taken by the macro-level institutions to articulate the community and local level units is thus worth investigating.

In the rural/offshore island sample (i.e., Kutubdia), all but two of the survey respondents received emergency assistance and/or relief (Table V). Such assistance from external sources or collective organizations was provided in several forms; the provision of food, medicine, tents and shelter, and cash-money were the main emergency relief measures, while corrugated iron sheets and seedlings were principal items supplied through rehabilitation programs. Due to extremely poor transportation and communication links with the mainland, and more importantly, within the island, relief supplies could not reach the more remote villages for many

Figure 6. A rural community in Kutubdia island devastated by the April 1991 cyclone. Storm surges and coastal flooding, along with high winds, caused the loss of myriad human lives; survivors had to also face the destruction of their cattle herds and subsistence livelihood. The relative effects of cyclones upon household economy in islands and rural aras thus can be catastrophic and far-reaching.

days. In some cases it took more than a week to detect and reach the affected areas. During this delays many survivors lived on only coconuts, particularly in the northern villages of the island. Although the entire affected population here eventually received some forms of relief assistance, the time lag created an acute problem for many survivors.

In the urban/mainland area (Bandar), more than three-quarters of the households received relief assistance. In a few cases in this locality, the affected but relatively well-off families refused relief to allow others access to assistance. Due to the well-developed road network and easy access, food and drinking water quickly reached most affected zones of the Bandar *thana.* Two types of institutions, specifically foreign relief organizations and various departments of the national government of Bangladesh, provided the most important emergency supports – such as food, medicine, and drinking water – to the victims. Notably, in the case of other types of natural hazards such as riverbank erosion and floods, it has been observed that relatives, friends, and other traditional social institutions play a predominant role in disaster recovery and rehabilitation (Haque, 1989; Haque and Zaman, 1991). Contrary to this finding, the role of traditional institutions was found to be minimal in the aftermath of the cyclone (Table V).

In rehabilitation measures, only two types of programs were undertaken by the prevailing national and regional governments and nongovernmental organizations. As noted above, one program offered help to reconstruct houses and the other provided inputs for agricultural rehabilitation. Once again, the involved institutions included both foreign relief organizations and local governments, the latter a lower-order arm of the national government.

A key question that arises here is the extent to which relief met the actual needs of the hazard victims. More than half of the respondents, both in Kutubdia and Bandar, indicated that more than three-quarters of their crucial emergency needs were fulfilled. Nonetheless, in the case of rehabilitation, a relatively smaller proportion of respondents were satisfied; only a nominal proportion of their requirements were provided by the macro-scale organizations. These patterns altogether support the postulation of this study that the prevailing institutions generally ignore the importance of rehabilitation.

In the present study, further attempts were made to substantiate the above hypothesis. Based on methodology drawn from perceptual and psychological studies, a number of questions were structured. Results from the 1992 survey regarding the two most relevant responses are presented in Table VI. The vast majority of the respondents, both in urban/mainland and rural/offshore islands stated that they did not receive any information from government departments regarding coping strategies with cyclones and relevant loss-mitigation measures (Table IV). A few respondents had been informed of the tree-planting program offered by the Department of Forestry; other programs had only been suggested to a nominal number of respondents. Nonetheless, implementation of tree-planting programs, in cooperation with the nongovernmental organizations, has been viewed as one of the desired actions among the rural inhabitants. However, only a very few respondents mentioned the need for improvement of nonstructural measures, such as information and communication facilities, emergency medical and health care, relief and rehabilitation, and land-use zoning.

5. Policy Implications and Recommendations

This study assessed the nature and characteristics of cyclone preparedness in Bangladesh at the institutional and individual level. The specific focus was given to the April 1991 cyclone warning process, the emergency preparedness and responses among coastal inhabitants, and the state of institutional actions to provide relief and rehabilitation assistance to the cyclone victims. The phenomenal magnitude of human losses and economic setbacks due to the disaster has generated serious adverse impacts on the resiliency of the affected people.

Since the general vulnerability of the coastal occupants to extreme environmental threats is not a new phenomenon, the local inhabitants are familiar with the cyclones and are quite aware of the related risks. Despite that awareness, lack of a choice to move elsewhere, combined with the relative economic benefits of

Table Va. Cross tabulation of post-cyclone assistance received by item and source, Bandar *thana*, Chittagong (urban) (percent; multiple response possible)

Source	Food	Tent/shelter	Medicine	Cash money	Seedlings	C.I. sheets
Relatives	11.9 (17)[a]	13.6 (3)	6.9 (6)	9.5 (2)	nil	nil
Friends	5.5 (8)	4.6 (1)	2.3 (2)	4.8 (1)	nil	nil
Local government	21.0 (30)	4.6 (1)	8.2 (7)	23.8 (5)	nil	9.4 (3)
National government	12.6 (18)	22.7 (5)	9.3 (8)	19.1 (4)	100.0 (1)	21.8 (7)
National relief agencies	10.5 (15)	9.1 (2)	11.6 (10)	4.8 (1)	nil	9.4 (93)
Foreign relief agencies	24.5 (35)	45.4 (10)	46.5 (40)	38.0 (8)	nil	43.8 (14)
Others	14.0 (20)	nil	15.2 (13)	nil	nil	15.6 (5)
Total	100.0 (143)	100.0 (22)	100.0 (86)	100.0 (21)	100.0 (1)	100.0 (32)

[a] Frequencies are shown in parentheses.
Source: Author's field survey, 1992.

Table Vb. Cross tabulation of post-cyclone assistance received by item and source, Kutubdia *thana*, Cox's Bazar (rural) (percent; multiple response possible)

Source	Food	Tent/shelter	Medicine	Cash money	Seedlings	C.I. sheets
Relatives	1.3 (2)[a]	nil	nil	nil	nil	2.0 (1)
Friends	nil	nil	nil	nil	nil	2.0 (1)
Local government	17.5 (27)	6.9 (2)	3.1 (3)	38.5 (10)	36.4 (16)	25.5 (13)
National government	35.1 (54)	3.5 (1)	22.5 (22)	19.2 (5)	15.9 (7)	9.8 (5)
National relief agencies	18.8 (29)	6.9 (2)	20.4 (20)	7.7 (2)	18.2 (8)	9.8 (5)
Foreign relief agencies	22.1 (34)	17.2 (5)	54.0 (53)	34.6 (9)	29.5 (13)	50.9 (26)
Others	5.2 (8)	65.5 (19)	nil	nil	nil	nil
Total	100.0 (154)	100.0 (29)	100.0 (98)	100.0 (26)	100.0 (44)	100.0 (51)

[a] Frequencies are shown in parentheses.
Source: Author's field survey, 1992.

living in the region, especially from the salt industry, fishing, and agriculture in and around the offshore islands, contribute to the dense rural settlement. Similarly, the favourable geographical sites for the seaport and heavy industries, and again fishing aid the economic and demographic growth of Chittagong and its environs.

In general, the disaster warning system in Bangladesh functions quite effectively in terms of identification, monitoring, and forecasting of meteorological conditions. Given resource constraints and the lack of infrastructural facilities in the country, the cyclone warning dissemination process in 1991 successfully alerted people to the potential threat and recommended emergency measures. Yet, a vast majority of the potential victims in coastal regions reacted passively to warnings due to disbelief, fear of burglars, and lack of proper cyclone shelters. The results of this study suggest that the cyclone warning system, in order to attain its goals, should regard the human response to warnings as its constituent part, and accommodate

Table VI. Distribution of recipients of precautionary suggestions from government sources since 1991 cyclone

Response	Bandar (urban/mainland)		Kutubdia (rural/offshore)	
	Frequency	%	Frequency	%
A. 'Have you received any suggestions on cyclone impact mitigation/rehabilitation measures from government departments?'				
Yes	15	9.3	17	16.0
No	147	90.7	89	84.0
	N = 162	%	N = 106	%
B. 'What type of suggestions have you received?' (multiple response possible)				
To plant more trees	13	8.0	17	16.0
To build *pucca* (brick/masonary) houses	5	3.1	2	1.9
To obtain a radio or television	7	4.3	6	5.7
Others	9	5.6	4	3.8

Source: Author's field survey, 1992.

these human dimensions in its operational design. It is recommended that the cyclone warning system must account for the response component, an analysis of which, together with the overall performance of the program, should form a basis for future strategies and policies. Rather than merely an isolated system, the hazard warning system should be regarded as a subset of a comprehensive disaster preparedness and management program.

In order to improve the rate and nature of response – that is, to make people more aware of the severity of the hazards and of precautionary options – educational and developmental schemes need to be formulated. Such plans however should be developed as part of income generating activities in order to help ameliorate the socioeconomic plight of the rural population. The study recommends that hazard mitigation policies should also be integrated with national economic development plans and programs. At the present time, the region has less than a thousand cyclone shelters, which together can accommodate less than a million people; the five million people vulnerable to cyclone damage all require access to such shelters. This expensive effort could be economized by constructing multipurpose shelters, buildings which could also serve as schools, health facilities and agricultural extension centers. Many communities in India have already adopted such a policy (Mishra and Prakash, 1982).

Resilience of the cyclone victims in sociopsychological terms is seen to be quite remarkable. But the perpetual process of impoverization and marginalization of the vast majority of the population is accelerated by the impact of environmental events. Due to the lack of appropriate rehabilitation, the ongoing process of cumulative vulnerability of these people will only cause more loss and damage in future hazard events. The trend of national and international resource allocation solely towards physical prevention and control of environmental events ignores the need to link the poor and marginalized disaster victims with the development process. The current system offers only partial solutions in disaster-mitigation efforts. It is therefore recommended that policy development that concerns disaster prevention and mitigation should integrate approaches from science and technology, and societal management.

Acknowledgements

This research was funded by the Brandon University Research Committee (BURC), Brandon University, Brandon, Canada, under the auspices of the Social Sciences and Humanities Research Council (SSHRC), Ottawa, Canada. The author is thankful to the BURC members for their support to this research, and to Ian Burton, James Boyce, Lolit Sinha, Gordon Winder, Glenn Bergen and two anonymous referees for their comments and criticisms on an earlier draft of the paper.

References

Anthes, R. A.: 1982, *Tropical Cyclones: Their Evolution, Structure and Effects*, Monograph Number 41. American Meteorological Society, Boston, Massachusetts.

BDRCS (Bangladesh Red Crescent Society): 1992, *A Guide to Volunteer Training. Cyclone Preparedness Program*, Dhaka (in Bengali).

Burton, I., Kates, R. W., and White, G. F.: 1993, *The Environment As Hazard,* 2nd edn, Guilford Press, New York.

Climate Analysis Center: 1991, *Weekly Climate Bull.* **91/18**, May 4, 1991, National Weather Service, NOAA, United States Department of Commerce, Washington, D.C.

Crossette, B.: 1991, Bangladesh toll now laid in part to poor aid plan, *The New York Times*, 4 May, New York.

Dube, S. K., Sinha, M. C., and Balasubramanyam, V.: 1982, The effect of coastal geometry on the location of peak surge, *Mausam* **33**, 445–450.

Emanuel, K. A.: 1988, Toward a general theory of hurricanes, *American Scientist* **76**, 371–379.

Frank, N. L. and Hossain, S. A.: 1971, The deadliest tropical cyclone in history?, *Bull. Amer. Meteorol. Soc.* **52**, 438–444.

Haque, C. E.: 1988, Impact of River-Bank Erosion in the Brahmaputra-Jamuna Floodplain: A Study of Population Displacement and Human Response Strategies, PhD Dissertation, University of Manitoba, Winnipeg.

Haque, C. E.: 1989, Human adjustments to river bank erosion hazard in the Jamuna Floodplain, Bangladesh, *Human Ecology* **16**, 421–437.

Haque, C. E.: 1991, Human responses to riverbank erosion hazard in Bangladesh: Some lessons from indigenous adjustment strategies, in K. M. Elahi *et al.* (eds), *Riverbank Erosion, Flood and Population Displacement in Bangladesh.* Jahangirnagar University, Dhaka, pp. 191–217.

Haque, C. E.: 1993, An appraisal of the disaster process in Bangladesh: What has been happening to the April cyclone victims?, in M. R. Wilson (ed.), *Proceedings of the Prairie Division,* Canadian Association Geographers, Department of Geography, University of Saskatchewan, Saskatoon, pp. 205–220.

Haque, C. E. and Blair, D.: 1992, Vulnerability to tropical cyclones: Evidence from the April 1991 cyclone in coastal Bangladesh, *Disasters* **16**, 217–229.

Haque, C. E. and Zaman, M. Q.: 1990, Coping with riverbank erosion hazard and displacement in Bangladesh: Survival strategies and adjustments, *Disasters* **13**, 309–323.
Haque, C. E. and Zaman, M. Q.: 1991, Working together: reflections on collaborative research in Bangladesh, *Canadian J. Development Studies* **12**, 387–403.
Haque, C. E. and Zaman, M. Q.: 1993, Human responses to riverine hazards in Bangladesh: A proposal for sustainable floodplain development, *World Development* **21**, 93–107.
Hastenrath, S. and Lamb, P. J.: 1979, *Climate Atlas of the Indian Ocean,* Part I: Surface Climate and Atmospheric Circulation, The University of Wisconsin Press, Madison, Wisconsin.
Hewitt, K.: 1983, The idea of calamity in a technological age, in K. Hewitt (ed.), *Interpretations of Calamity from the Viewpoint of Human Ecology,* Allen and Unwin, London, pp. 3–32.
Hewitt, K. and Burton, I.: 1971, *The Hazardousness of a Place: A Regional Ecology of Damaging Events*, Department of Geography, University of Toronto, Toronto.
Islam. M. A.: 1971, *Human Adjustment to Cyclone Hazards: A Case Study of Char Jabbar*, Natural Hazards Research Working Paper No. 18. University of Toronto, Toronto.
Islam, M. A.: 1974, Tropical cyclone, coastal Bangladesh, in G. F. White (ed.), *Natural Hazards: Local, National, Global*, Oxford University Press, New York, pp. 19–24.
Islam, M. S. *et al.*: 1992, A case study on April 1991 cyclone of Bangladesh, A paper presented at the workshop on Improving Cyclone Warning Response and Mitigation, January 19–30, 1992, Dhaka.
Ittelson, W. H. *et al.*: 1974, *An Introduction to Environmental Psychology*, Holt, Rinehart and Winston, New York.
Johns, B., Sinha, Dube, P. C., Dube, Mohanty, U. C., and Rao, A. D.: 1983, On the effect of bathymetry in numerical storm surge simulation experiments, *Computers and Fluids* **11**, 161–174.
Khalil, G. M.: 1992, Cyclones and storm surges in Bangladesh: Some mitigative measures, *Natural Hazards* **6**, 11–24.
Mamun, M. Z.: 1996, Awareness, preparedness and adjustment measures of river-bank erosion-prone people: A case study, *Disasters* **20**, 68–74.
Mirza, M. M. K. and Pal, S.: 1992, *Prakritik Durjog O Bangladesher Poribesh,* Centre for Environmental Studies and Research, Dhaka (in Bengali).
Mishra, D. K. and Prakash, H. R.: 1982, An evaluation of the Andhra Pradesh cyclone shelters programme, *Disasters* **6**, 250–255.
Mooley, D. A.: 1980, Severe cyclonic storms in the Bay of Bengal, 1877–1977, *Monthly Weather Review* **108**, 1647–1655.
Palm, R.: 1990, *Natural Hazards: An Integrative Framework for Research and Planning,* John Hopkins University Press, Baltimore and London.
Paul, B. K.: 1992, Farmers' response to drought in Bangladesh: Lessons for planners and administrators, Paper presented at the Annual Meeting of the Association of American Geographers, San Diego, April 18–22.
Raghavendra, V. K.: 1973, A statistical analysis of the number of tropical storms and depressions in the Bay of Bengal during 1890–1969, *Indian J. Meteorol. Geophys.* **24**, 125–130.
Rahman, M. H.: 1995, Responding to drought in Bangladesh, *The Daily Star* 15 May, 6–10.
Rasid, H. and Paul, B. K.: 1984, Perception of and adjustment to floods in Jamuna floodplain, Bangladesh, *Human Ecology* **12**, 3–19.
Sevenhuysen, G. P.: 1991, Report on cyclone disaster response in Bangladesh, Disaster Research Unit, University of Manitoba, Winnipeg, Canada.
Sims, J. H. and Baumann, D. D.: 1983, Educational programs and human response to natural hazards, *Environment and Behaviour* **15**, 165–189.
Sinha, A. K. and Avrani, S. U.: 1984, 'The disaster warning process: A study of the 1981 Gujarat cyclone', *Disasters* **8**, 67–73.
Smith, K.: 1992, *Environmental Hazards: Assessing Risk and Reducing Disaster*, Routledge, New York.
Warrick, R. A. *et al.*: 1981, *Four Communities Under Ash,* Monograph Number 34, Institute of Behavioral Science, University of Colorado, Boulder.

Williams, H. B.: 1964, Human factors in warning and response systems, in G. Grosser *et al.* (eds), *The Threat of Impending Disaster: Contributions to the Psychology of Stress*, MIT Press, Cambridge, Massachusetts, pp. 75–83.

Natural Hazards **16:** 203–218, 1997.

Estimation of Response Spectra in the Severely Damaged Area During the 1995 Hyogo-Ken Nambu Earthquake

TADANOBU SATO[1]
[1]*Disaster Prevention Research Institute, Kyoto University, Uji, Kyoto, Japan*

KATSUTOSHI KITA[2]
[2]*Department of Ocean Civil Engineering, Tokai University, Shimizu, Shizuoka, Japan*

TOMONARI MAEDA[3]
[3]*Department of Civil Engineering, University of Tokyo, Tokyo, Japan*

(Recieved: 18 December 1996; in final form: 14 August 1997)

Abstract. We calculated theoretical acceleration response spectra at the ground surface in the region near the 1995 Hyogo-ken Nambu earthquake source based on the spectral moment method. To estimate earthquake motion on the ground surface, a formula of earthquake motion at base rock level was derived. The amplification effect of the ground was introduced by using multiple reflection theory. Theoretically estimated response spectrum were modified by the response spectra calculated using observed earthquake motions.

Key words: response spectra, earthquake source mechanism, wave propagation, amplification of ground motion

1. Introduction

The 1995 Hyogo-ken Nambu earthquake ($M = 7.2$) struck the densely populated corridor between the cities of Osaka and Kobe at 5:46 am local time, Tuesday, 17 January 1995. As of 29 December 1995, the earthquake resulted in 6308 deaths and 43 177 persons injured. The respective numbers of completely and partially collapsed dwellings are 100 302 and 366 114. The estimated cost of damage is 10 trillion yen: 5.8 trillion yen to buildings, 1.4 trillion yen to port and harbors, 450 billion yen to highways, 420 billion yen to gas and power supply systems, and 340 billion yen to railways.

The earthquake generated a large number of strong motion recordings over a wide variety of geological site conditions, including rock and soil, as well as in structures of varying types. Several agencies, such as the Committee of Earthquake Observation and Research in the Kansai Area (CEORKA), Osaka Gas, Kansai Electricity, Japan Railway, and the Building Research Institute of the Ministry of Construction, maintain strong motion instrumentation networks. A plot of the peak horizontal ground acceleration vs. the shortest distance from the fault zone of the

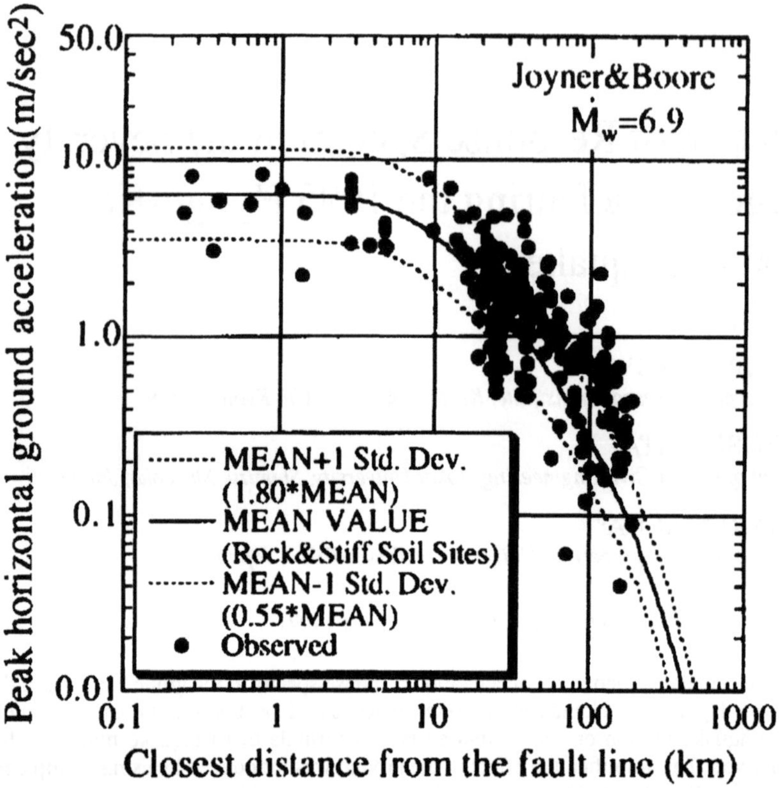

Figure 1. Relationship between peak acceleration and shortest distance to fault (Ejiri *et al.*, 1995). Thick line: Attenuation relationship proposed by Joyner and Boore (1981); Broken lines: Standard deviation $\sigma = 0.1$.

mainshock is shown in Figure 1 (Ejiri *et al.*, 1995). This is referred to the mean attenuation relationship for hard soil by Joyner and Boore (1981), for a moment magnitude of 6.9.

The cities of Kobe, Ashiya, and Nishinomiya are located between the Rokko Mountains and Osaka Bay. The ground motions are believed to be amplified in the plain at the base of the Rokko Mountains. Figure 2 shows the distribution of active faults and the area assigned the seismic intensity of 7 (JMA scale), defined as the zone in which the collapse of wooden houses was more than 30%. The severely damaged area is 1–2 km to the south and parallel to the active faults and has a width of 2 km. The acceleration and velocity response spectra for the Kobe Maritime Meteorological Observatory and other sites are shown in Figure 3 (Ejiri *et al.*, 1995). Maximum response acceleration exceeds 2G for structures with a predominant period of 0.3–0.9 sec. A comparison with the design spectra for level-2 seismic force defined in the code of earthquake resistant design of bridges (1990) with actual response spectra is shown in Figure 4 (Ejiri *et al.*, 1995). The actual spectra for hard ground exceeds 2G which is almost twice the level of the

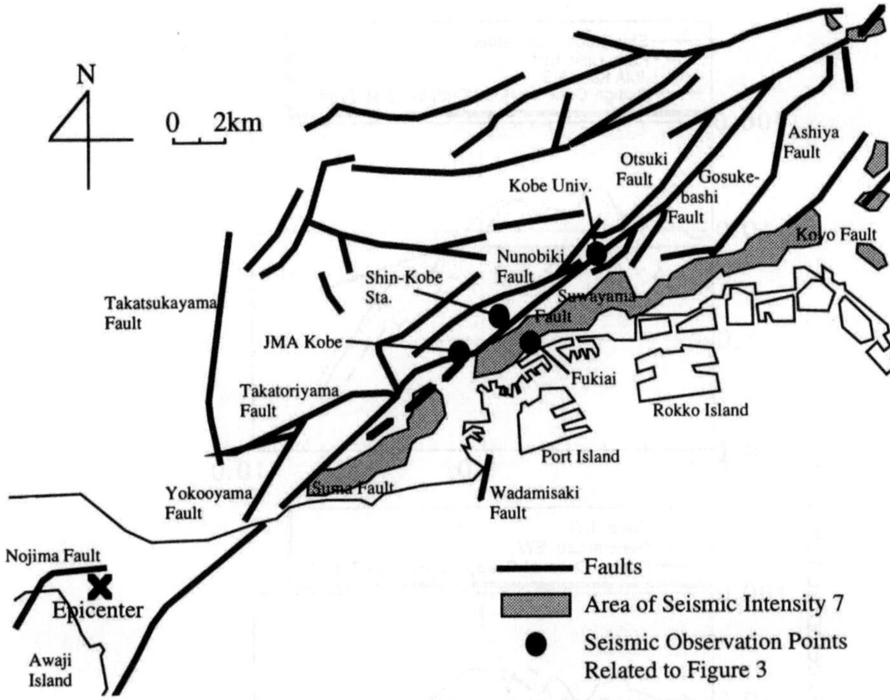

Figure 2. Map of faults and area of seismic intensity 7 (Suetomi, 1995).

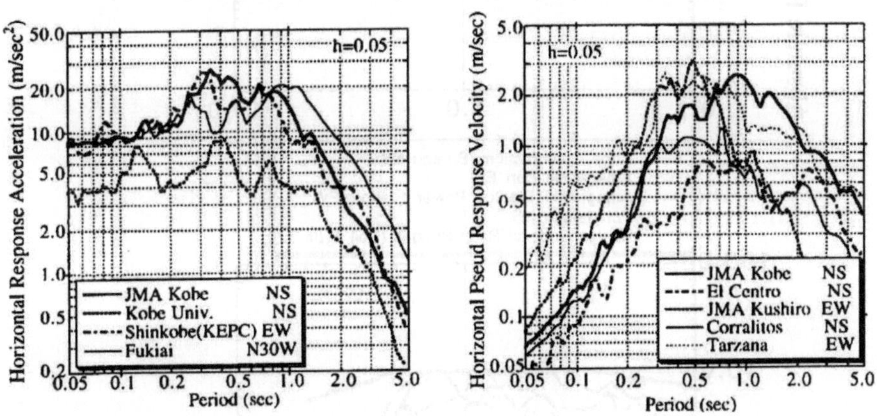

Figure 3. Comparison of acceleration and velocity response spectra (Ejiri *et al.*, 1995).

design spectrum. The design spectrum for soft ground is almost the envelope of the real response spectra.

We estimate response spectra to investigate the dynamic response characteristics of structures located in the severely damaged area. The response of a single degree of freedom system (SDF) in the frequency domain is derived by multiplying the strong motion at the base rock level with the transfer function of the soil and the

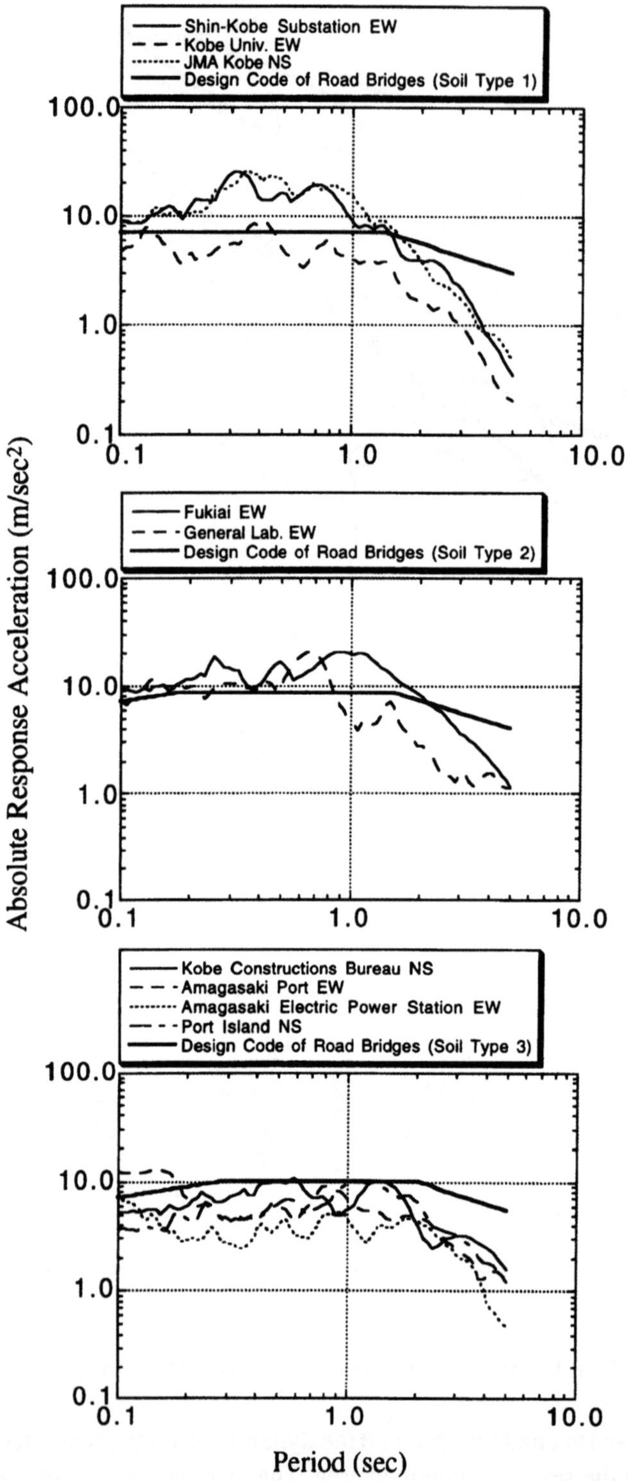

Figure 4. Comparison of acceleration response spectra calculated using observed earthquake motions and design spectra of the level-2 seismic force (Ejiri *et al.*, 1995).

frequency transfer function for a single degree of freedom system. A simplified analytical method is introduced which predicts earthquake motion at base rock level near the source region, where the effect of the fault dimension cannot be ignored. For this purpose, a large event is assumed to be synthesized from small events on the fault plane. This method was first proposed by Hartzell in a simple formula, and others have attempted to improve his method (Hartzell, 1978; Muramatsu and Irikura, 1981). We used Irikura's revised model (Irikura, 1983), which is based on the idea that slips on the fault plane during a large event can be replaced by the spatial distribution of slips from small events. An ω^{-2} scaling is used in the high frequency range.

Site effects are important when predicting earthquake motion on the surface. Assuming a shear wave velocity of 3.4 km/sec in the basement, the amplification effect at the surface is evaluated from the transfer function using multiple reflection theory. The damping effect through the propagation path is also taken into account.

2. Analytical Procedure for Estimating Response Spectra

2.1. ESTIMATION OF THE EARTHQUAKE MOTION AT THE BASEMENT LEVEL

A fault model of the Haskell type is assumed. The model is described by a rectangular fault with fault length, L; fault width, W; rise time, τ; final dislocation, D; and rupture velocity, v_r. A large event can be synthesized by superposing small events and taking into account the time delay caused by the rupture propagation.

The fault plane is divided into n elements, each of which corresponds to the area of a small event. The motion of the large event, $g_{OL}(t)$, at the observation point is expressed by (Irikura, 1983)

$$g_{OL}(t) = \sum_{i=1}^{n_L}\sum_{j=1}^{n_W} g_{OS}(t - t_{ij}) + \sum_{i=1}^{n_L}\sum_{j=1}^{n_W}\sum_{k=1}^{(n_D-1)n'} g_{OS}(t - t_{ijk}), \quad (1)$$

where $g_{OS}(t)$ is a small event, n_L, n_W, and n_D are the number of subdivisions corresponding to the fault length, fault width, and dislocation, and t_{ij} and t_{ijk} are time delays expressed by

$$t_{ij} = \frac{(R_{ij} - R_0)}{v_s} + \frac{\xi_{ij}}{v_r}, \qquad t_{ijk} = t_{ij} + \frac{k\tau}{(n_D - 1)n'}, \quad (2)$$

in which R_0 is the distance between the observation point and the hypocenter, R_{ij} the distance between a point on the basement and the initial point of rupture at ij element, ξ_{ij} the distance between the hypocenter and the initial point of the rupture at ij element, v_s the shear wave velocity of the medium, v_r the rupture velocity, τ the rise time, and n' is an arbitrary integer to eliminate aliasing. The following relation is assumed for the number of superposition

$$n_W = n_L = n_D = n = \left(\frac{M_{OL}}{M_{OS}}\right)^{1/3} \quad (3)$$

in which M_{0L} and M_{0S} are seismic moments for the large and small events, respectively. In the following we assume $n' = n = 3$.

The Fourier transform of Equation (1) is

$$g_{0L}(\omega) = \sum_{i=1}^{n_L}\sum_{j=1}^{n_W} g_{0S}(\omega)\exp(-i\omega t_{ij}) + \\ + \sum_{i=1}^{n_L}\sum_{j=1}^{n_W}\sum_{k=1}^{(n_D-1)n'} g_{0S}(\omega)\exp(-i\omega t_{ijk}) \tag{4}$$

in which $g_{0S}(\omega)$ is the Fourier spectrum of a small event given by

$$g_{0S}(\omega) = \frac{R_{\theta\phi}}{4\pi\rho v_s^3}\frac{s(\omega)}{R_{ij}}, \tag{5}$$

in which $R_{\theta\phi}$ is the radiation pattern, ρ the medium density, R_{ij} the distance from a fault element to the site, and $s(\omega)$ the source spectrum. We adopt the formula proposed by Geller (1975) for the source spectrum of each small event with ω^{-2} rollout in the high frequency range. Because the number of superposition of the second term on the right-hand side of Equation (4) is n^3 its slope in the high frequency range is ω^{-3}. The first term on the right-hand side of Equation (4), however, has a logarithmic slope of -2 in the high frequency range, hence $g_{0L}(\omega)$ shows ω^{-2} characteristic.

From Irikura's revised model, the source displacement spectrum, $g_{0L}(\omega)$, has a logarithmic slope of -2 in the high frequency range. The source spectrum of acceleration is flat in that range. To prevent the integral of the second-order spectral moment from becoming infinite we introduce $f_{\max}$, the corner frequency of the acceleration spectrum in the high frequency range. According to Hanks (1982), the spectrum drops off sharply for frequencies between 20–30 Hz on a rock site. A value of $f_{\max}$ of 25 Hz was selected.

2.2. PATH EFFECT

In Equation (5) only geometrical spreading proportional to the reciprocal distance, $1/R$, is considered. S-wave damping may be estimated by (Aki and Richards, 1980)

$$g_S(\omega) = g_{0S}(\omega)\exp\left(-\frac{\omega R_{ij}}{2v_s Q}\right), \tag{6}$$

where $g_S(\omega)$ is the Fourier transform of ground motion. The Q-value for the Kinki region (Akamatsu, 1980) is approximated by

$$\log Q^{-1} = -0.5\log\frac{\omega}{2\pi} - 2.3 \tag{7}$$

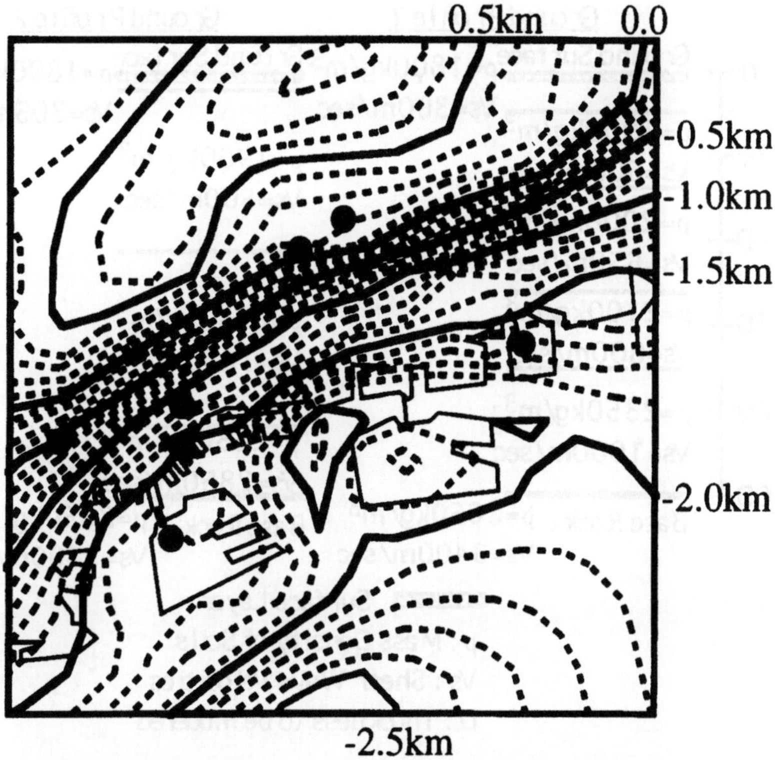

Figure 5. Configuration of top surface of the base rock in the Kobe sedimentary basin proposed by Nakagawa *et al.* (1996).

in which ω is an angular frequency.

2.3. AMPLIFICATION EFFECT

To estimate the amplification we collected the standard penetration test data of soil profiles in the Kobe area and we drew the contours of $N = 50$. The frequency transfer function in the near-surface soils, $H_s(\omega)$, is calculated based on the multiple reflection theory assuming horizontally layered soil. The ground motion at the surface, $g_L(\omega)$, is expressed by

$$g_L(\omega) = H_s(\omega)H_d(\omega)g_{0L}(\omega) \tag{8}$$

in which $H_d(\omega)$ is the frequency transfer function from the base rock with a shear wave velocity of 3.4 km/sec to the soil layer with the N-value of 50. Nakagawa *et al.* (1996) determined the depth to the base rock in the Kobe area by use of the Bouguer anomaly as shown in Figure 5. The geological profile above this base rock is assumed to be similar to one of the profiles shown in Figure 6. When the

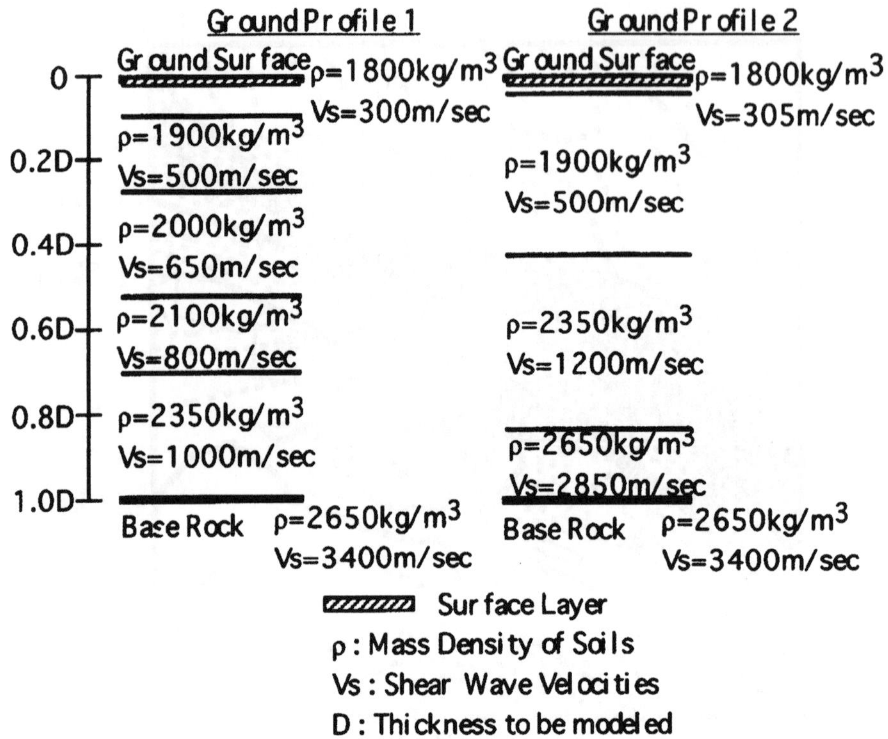

Figure 6. Horizontally layered ground models for deeper parts of ground profiles.

information of the deeper part of ground profile is not available the value of $H_d(\omega)$ is assumed to be 1.75 as proposed by Midorikawa and Kobayashi (1988).

2.4. RESPONSE SPECTRUM

The response of a single degree of freedom system at the ground surface $X(\omega)$ is expressed by multiplying the frequency transfer function of SDF, $H_x(\omega, T_x, h_x)$ with $g_L(\omega)$ as follows

$$X(\omega) = H_x(\omega, T_x, h_x) H_s(\omega) H_d(\omega) g_{0L}(\omega) \tag{9}$$

in which T_x is the natural frequency of SDF and h_x is the damping constant. The power spectrum of the response of SDF is

$$P(\omega) = \frac{1}{T} X(\omega) X^*(\omega) \tag{10}$$

in which T is the duration of the stationary part of the ground motion and the asterisk indicates the complex conjugate.

The duration, T, is defined as

$$T = T_e - T_b + T_\alpha, \tag{11}$$

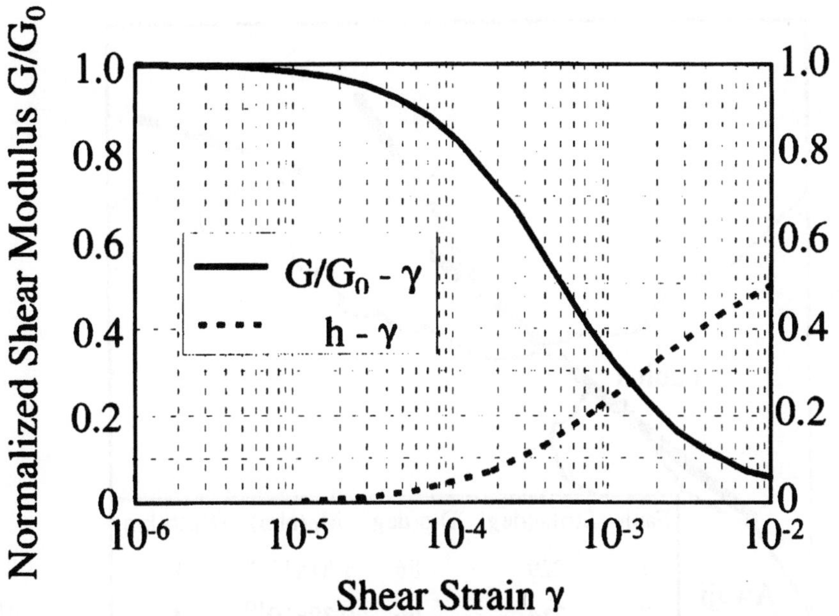

Figure 7. An example of shear strain dependence of equivalent shear modulus and damping constant.

in which T_b and T_e are the arrival times of the seismic waves originating from the initial and final rupture elements, and T_α is the duration of the seismic wave generated in the last rupture element (Kawashima *et al.*, 1984).

Once the power spectrum of the large event at the rock site has been obtained, the expected absolute peak acceleration response of the SDF, $S_a(T_x, h_x)$, is calculated as follows (Der Kiureghian, 1980):

$$S_a(T_x, h_x) = p\sqrt{\lambda_0}, \quad p = f(\nu_e, \lambda_0, \lambda_1, \lambda_2) \tag{12}$$

in which p is the peak coefficient, ν_e the reduced ratio of the zero-crossing given as a function of the three spectral moments and the duration, and λ_m (m = 0, 1, 2) are the spectral moments of zero and the first and second orders given by

$$\lambda_m = \int_0^\infty \omega^m P_A(\omega)\mathrm{d}\omega. \tag{13}$$

where $P_A(\omega) = \omega^4 P(\omega)$ is the power spectrum of the acceleration resonse. If soil softening occurs we must consider the nonlinear response of the soil layer in the estimated frequency transfer function. We use a Hardin–Drnevich type relationship between the equivalent shear strain and the equivalent shear modulus as well as the damping constant as shown in Figure 7. The equivalent shear strain is assumed to be 0.65 times the peak shear strain at the center of each layer. The peak shear strain is calculated through the same procedure to estimate the peak acceleration

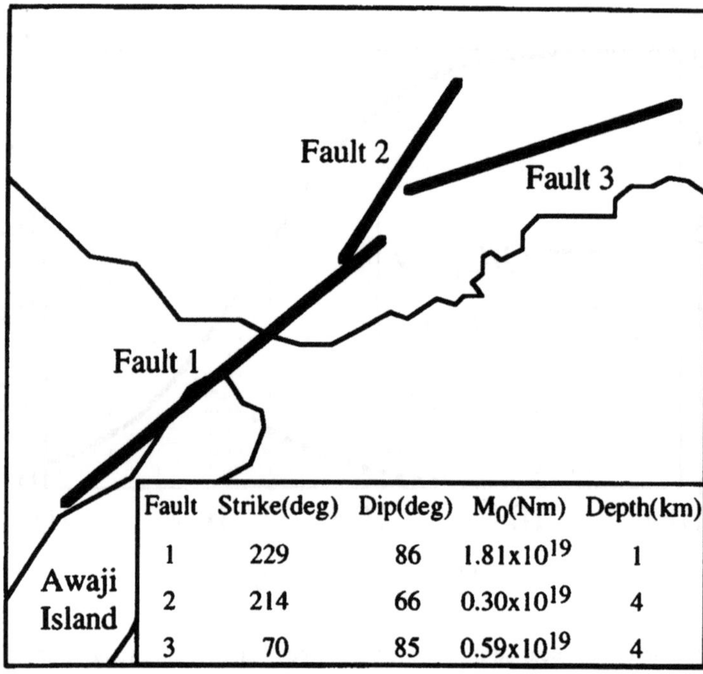

Fault	Strike(deg)	Dip(deg)	M_0(Nm)	Depth(km)
1	229	86	1.81×10^{19}	1
2	214	66	0.30×10^{19}	4
3	70	85	0.59×10^{19}	4

Figure 8. Locations of three subfaults and source parameters determined by using Kikuchi's source mechanism.

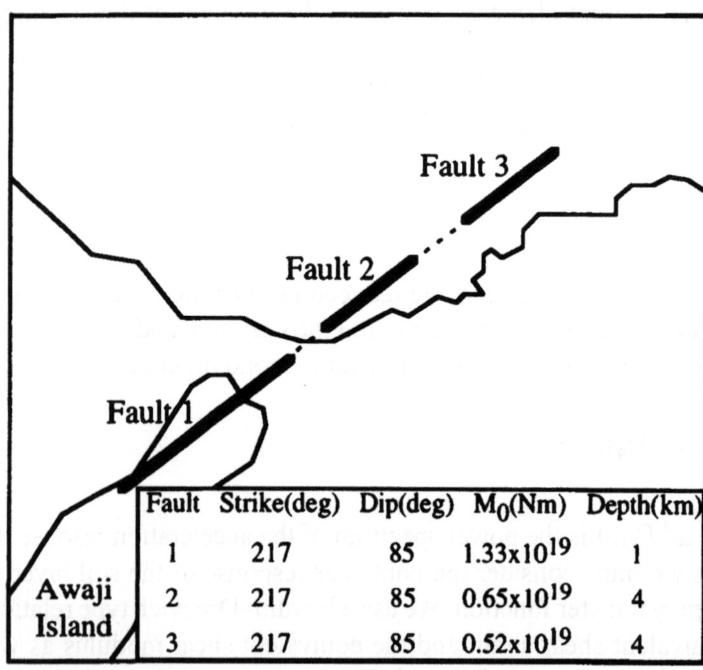

Fault	Strike(deg)	Dip(deg)	M_0(Nm)	Depth(km)
1	217	85	1.33×10^{19}	1
2	217	85	0.65×10^{19}	4
3	217	85	0.52×10^{19}	4

Figure 9. Locations of three subfaults and source parameters determined by using Irikura's source mechanism.

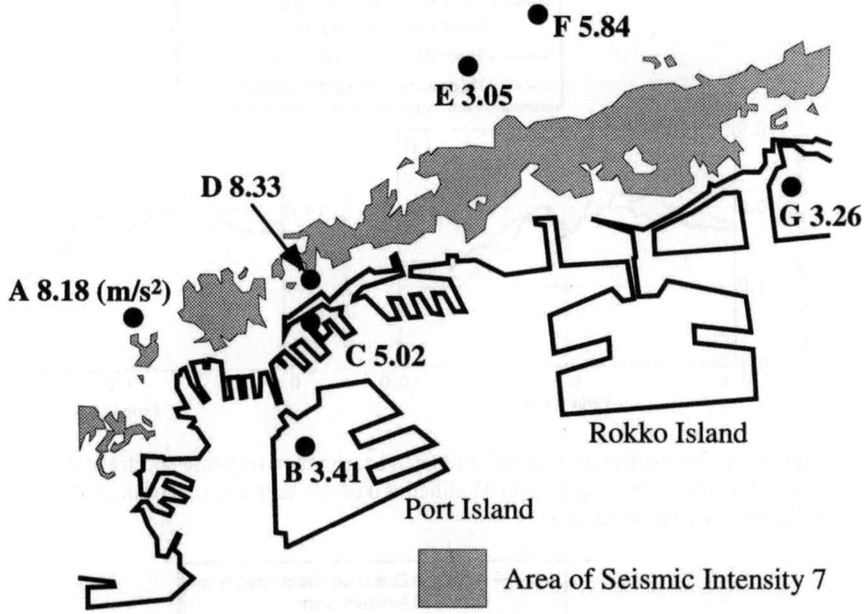

Figure 10. Locations of observation sites and peak accelerations at observation sites.

for Equation (12). Iterative calculation of the peak strain is confirmed to reach the converging error of strain less than 1% of the peak strain at the previous iteration step.

3. Estimation of Response Spectra for Kobe Earthquake

Equation (4) can be extended to multiple fault rupture mechanisms. The 1995 Hyogo-ken Nambu Earthquake was composed of three sub-events based on Kikuchi and Irikura (Kikuchi, 1995; Pitarka *et al.*, (1995). Kikuchi's model is composed of three point source sub-events obtained from inversion of teleseismic body waves. We assume the seismic moments of the first, second and third subfaults to be $M_0 = 1.81 \times 10^{19}$, 0.30×10^{19} and 0.59×10^{19} Nm (Kikuchi, 1995). The lengths of the first, second, and third subfaults are assumed to be 29, 16 and 20 km. The ratio of the length to the width of each subfault is assumed to be 0.5. The strike and dip angles for each subfault are given in Figure 8. The locations of the three subfaults are shown in Figure 8 (Kikuchi, 1995). We subdivided each subfault into nine subelements of equal area. Rupture of the first subfault is assumed to start at the bottom center element, and the ruptures of the second and third subfaults at the bottom southwest corner element. Two seconds after the start of the first rupture the second subfault ruptures and two seconds later the third rupture is assumed to occur. Pitarka *et al.* (1995) propose a source rupture mechanism to explain the short period ground motion near the source region. We assume three fault planes

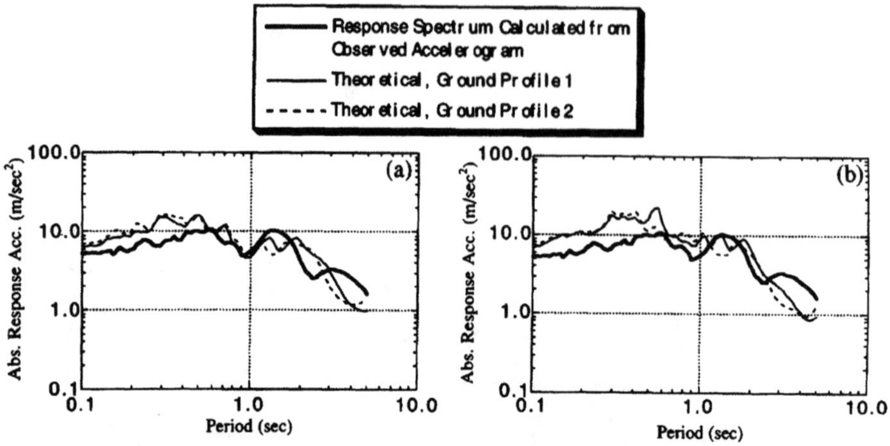

Figure 11. Comparison of real and estimated acceleration response spectra at the observation point C for the case of using; (a) Modified Kikuchi's source mechanism, and; (b) Modified Irikura's source mechanism.

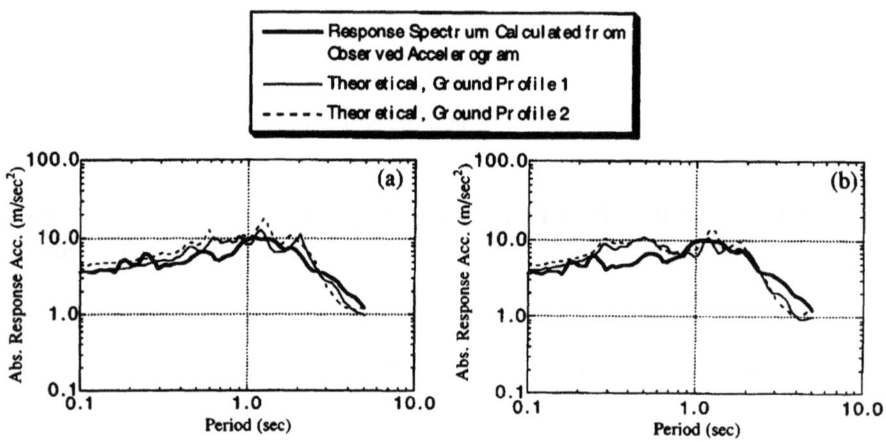

Figure 12. Comparison of real and estimated acceleration response spectra at the observation point B for the case of using (a) Modified Kikuchi's source mechanism, and (b) Modified Irikura's source mechanism.

on the basis of their solution. The strike and dip angles as well as seismic moment are given in Figure 9. The location of three subfaults are also shown in Figure 9. The length of the first, second, and third subfaults are assumed to be 15, 7.5 and 7.5 km (Pitarka *et al.*, 1995). The ratio of the length to the width of each subfault is assumed to be 0.5. We subdivided each subfault into nine subelements of equal area. Rupture of the first subfault is assumed to start at the northeast of the bottom corner element, and the ruptures of the second and third subfaults from the bottom southwest corner element. Two seconds after the start of the first rupture the second subfault ruptures and two seconds later the third rupture is assumed to occur.

The locations of seismic observation points are shown by A to G in Figure 10, together with the observed peak accelerations at these points. At each site we

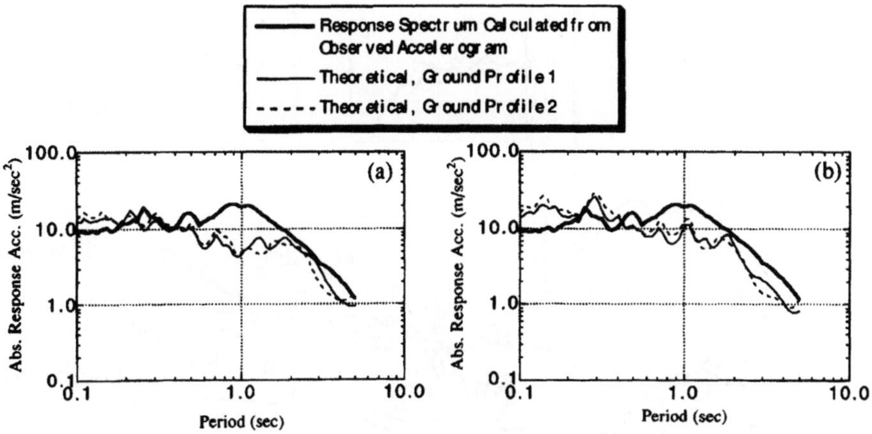

Figure 13. Comparison of real and estimated acceleration response spectra at the observation point D for the case of using (a) Modified Kikuchi's source mechanism, and (b) Modified Irikura's source mechanism.

calculate the Fourier spectra of surface ground motion by using Equation (8) to evaluate the amplification effect. Seven records were available for the calculation of response spectra in Kobe city. The estimated acceleration response spectra are compared with actual response spectra from records. Figures 11a and b show the results obtained using Kikuchi's and Irikura's fault rupture models at point C. The heavy, thin and broken lines are respectively the real response spectrum and the estimated response spectra using soil profiles of type 1 and type 2 (Figure 6). There is not much difference between real response spectra and estimated ones except in the period range from 0.1 to 0.5 sec. The estimated response spectra are very similar even though we assume fairly different fault rupture mechanisms and ground profiles. At point B where liquefaction occurred we can see very good agreement between the real and estimated absolute response spectra (Figure 12). Thus equivalent linear analysis provides an efficient tool for estimating the seismic response of liquefied ground. The comparison of real and estimated response spectra at point D is shown in Figure 13. This point is located near the south edge of the belt of seismic intensity 7 where the irregular profile of deeper part of ground structure is said to play a predominant effect on the amplification of ground motion in the period range of 0.5 to 2 sec (Irikura, 1995). Fairly large discrepancies can be seen between real and estimated spectra especially when we use Kikuchi's fault rupture model. For Irikura's fault rupture model the level of the estimated response spectrum is lower than the real one in the period range higher than 0.5 sec and the opposite can be seen in the period range lower than 0.5 sec. Schematic distribution of absolute acceleration response spectra obtained by using Irikura's fault rupture model and type 2 ground profile is given in Figure 14 with the belt of seismic intensity 7.

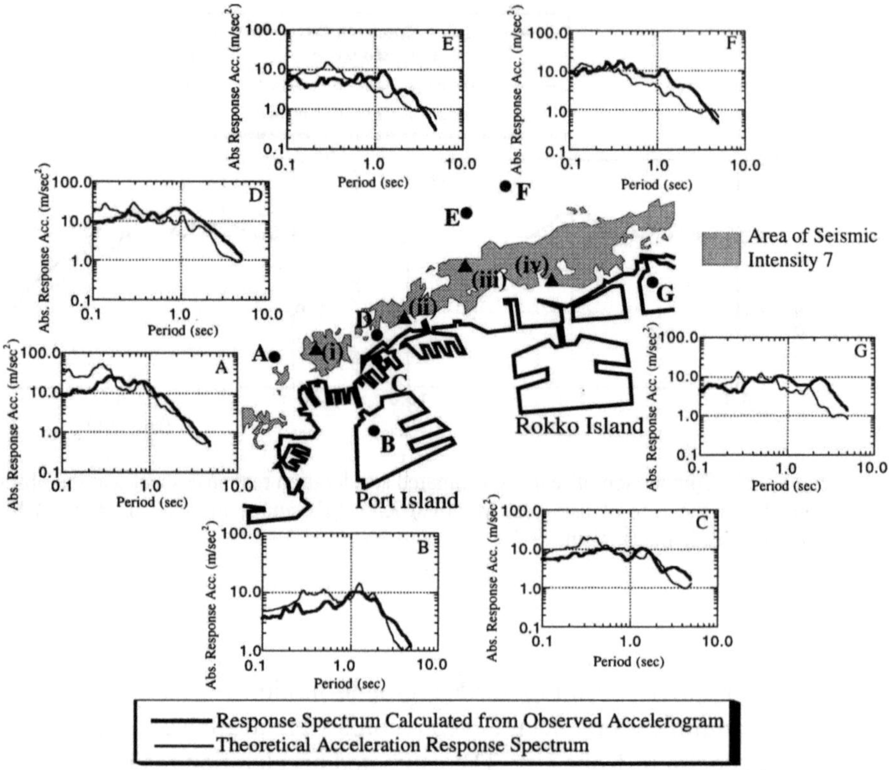

Figure 14. Schematic distribution of estimated acceleration response spectra at seven observation points.

4. Modification of Estimated Response Spectra

The estimated acceleration response spectra generally are in good agreement with the observed response spectra, but there are some discrepancies. We may improve the agreement as follows. Consider three observation points $(x_m, y_m)(m = i, j, k)$. Let $f_a(x, y, T_0)$ be the theoretically estimated peak absolute acceleration response of SDF with natural period T_0. The real peak absolute response of SDF at (x_m, y_m) is $A_m(T_0)$. Then the ratio between estimated and real response of SDF is given by

$$\log\{\Delta A_m(T_0)\} = \log\{A_m(T_0)\} - \log\{f_a(x_m, y_m, T_0)\}. \quad (14)$$

The interpolated peak absolute acceleration response of SDF with natural period T_0 at some point (x, y) inside the triangle composed of three observation points is expressed as

$$\log\{f_e(x, y, T_0)\} = \log\{f_a(x, y, T_0)\} + \log\{\Delta A(x, y, T_0)\}, \quad (15)$$

in which $\Delta A(x, y, T_0)$ is interpolated from $A_m(T_0)(m = i, j, k)$ using a linear interpolation function as follows:

$$\log\{\Delta A(x, y, T_0)\} = g(\Delta A_i, \Delta A_j, \Delta A_k, x, y). \quad (16)$$

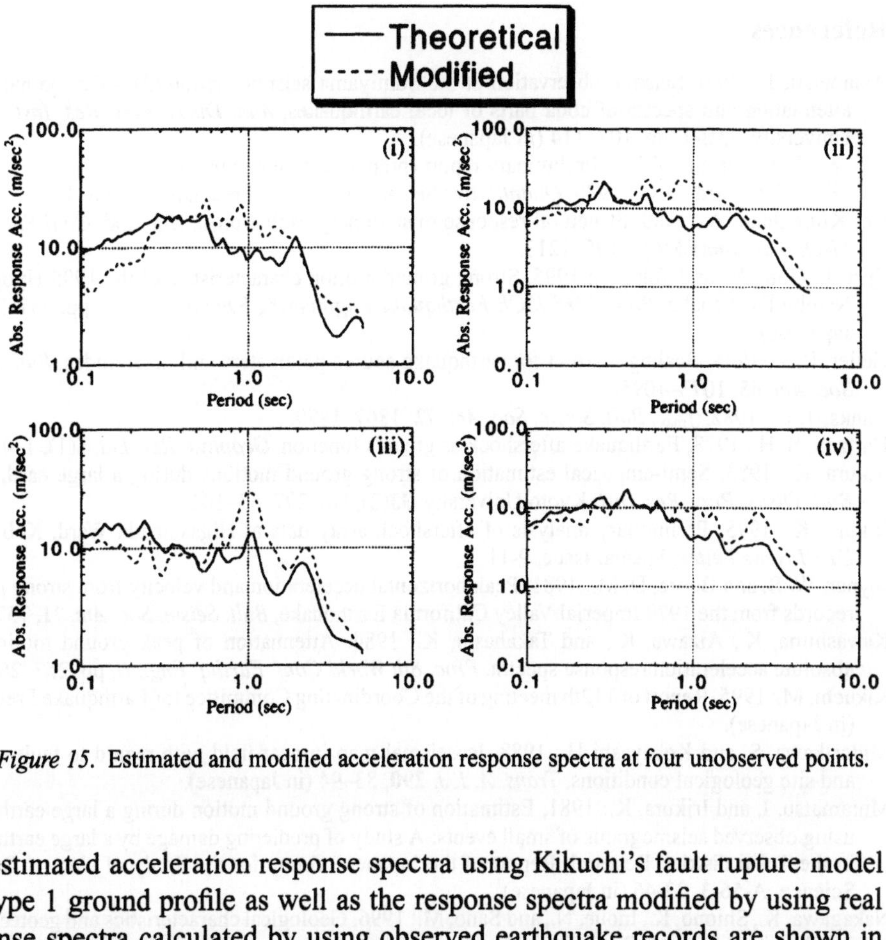

Figure 15. Estimated and modified acceleration response spectra at four unobserved points.

The estimated acceleration response spectra using Kikuchi's fault rupture model and type 1 ground profile as well as the response spectra modified by using real response spectra calculated by using observed earthquake records are shown in Figure 15. Point (i) is near Sannomiya station of Hankyu Railway; point (ii) at a damaged framed pier of Hanshin Expressway; point (iii) at the Rokko-michi station of JR; and point (iv) at a damaged single column RC bridge pier (each point is shown in Figure 14 by triangles). Note that the structure response reached 2G inside the belt with seismic intensity 7.

5. Conclusions

Acceleration response spectra at ground surface were calculated theoretically for the 1995 Hyogo-ken Nambu Earthquake, including the effect of fault extent, path effect and amplification due to the shallow and deep parts of ground profiles. The estimated acceleration response spectra at observation sites were compared with the real spectra obtained from recorded earthquake motions. The estimated response spectra at sites without observations were modified by using the real response spectra at observation sites. At several places the level of the estimated acceleration response exceeded 2G in the period range from 0.2 to 1.5 sec.

References

Akamatsu, J.: 1980, Seismic observation at the Sumiyama seismic station (2) – On the natures of attenuation and spectra of coda parts of local earthquakes, *Ann. Disas. Prev. Res. Inst.*, Kyoto University, 23B-1, pp. 107–114 (in Japanese).

Aki, K.: 1984, Origin of f_{max}, Preliminary report (prepublication manuscript).

Aki, K. and Richards, P.G.: 1980, *Quantitative Seismology, Vol. 1.* Freeman, pp. 168–169.

Der Kiureghian, A.: 1980, Structural response to stationary excitation, *J. Am. Soc. Civil Eng., Eng. Mech. Div.* **106**(EM6), 1195–1213.

Ejiri, J., Goto, Y., and Toki, K.: 1995, Strong ground motion characteristics of the 1995 Hyogo-ken Nambu Earthquake, *Proc. 23rd JSCE Earthquake Engineering Sympos. – 1995*, pp. 237–240 (in Japanese).

Geller, R. J.: 1975, Scaling relation for earthquake source parameters and magnitudes, *Bull. Seism. Soc. Am.* **65**, 1073–1095.

Hanks, T. C.: 1982, f_{max}, *Bull. Seism. Soc. Am.* **72**, 1867–1879.

Hartzell, S. H.: 1978, Earthquake aftershock as green's function, *Geophys. Res. Let.* **5**(1), 1–4.

Irikura, K.: 1983, Semi-empirical estimation of strong ground motions during a large earthquake, *Bull. Disas. Prev. Res. Inst.* Kyoto University, **33**(2), No. 297, 63–104.

Irikura, K.: 1995, Preliminary analysis of aftershock array data at Higashinada Ward, Kobe City, *DPRI News Letter*, Special Issue, 9-11.

Joyner, W. B. and Boore, D. M.: 1981, Peak horizontal acceleration and velocity from strong motion records from the 1979 Imperial Valley California Earthquake, *Bull. Seism. Soc. Am.* **71**, 757–783.

Kawashima, K., Aizawa, K., and Takahashi, K.: 1984, Attenuation of peak ground motion and absolute acceleration response spectra, *Proc. 8th World Conf. Earthq. Eng., II*, pp. 257–264.

Kikuchi, M.: 1995, Report of 112th meeting of the Coordinating Committee for Earthquake Prediction (in Japanese).

Midorikawa, S. and Kobayashi H.: 1988, Isoseismal map in near-field with regard to fault rupture and site geological conditions, *Trans. A. I. J.* **290**, 83–94 (in Japanese).

Muramatsu, I. and Irikura, K.: 1981, Estimation of strong ground motion during a large earthquake using observed seismograms of small events: A study of predicting damage by a large earthquake in the Tokai district, Research Report of the Japanese Group for the Study of Natural Disaster Science, A-56-3, 33-46 (in Japanese).

Nakagawa, K., Shiono, K., Inoue, N., and Sano, M.: 1996, Geological characteristics and geotechnical problem in and around Osaka Basin for basis to assess the seismic hazards, *Soils and Foundations*, Special Issue for the 1995 Hyogo-ken Nambu Earthquake, pp. 15–28.

Pitarka, A., Irikura, K., and Kagawa, T.: 1995, Source complexity of the January 17, 1995 Hyogoken-nambu earthquake determined by near-field strong motion modeling, preliminary results, *J. Natural Disaster Sci.* **16**, 31–38.

Suetomi, I.: 1995, Personal Communication.

Natural Hazards **16:** 219–241, 1997.

Federal Legislation for Disaster Mitigation: A Comparative Assessment Between Canada and the United States*

JOHN NEWTON
John Newton Associates, 262 Robert Street, Toronto, Canada M5S 2K8

(Received: 30 December 1996; in final form: 26 June 1997)

Abstract. Exposure to natural hazards tends to be poorly understood and yet implicitly accepted by a large proportion of populations at risk. Some people through years of experience and time-tested coping mechanisms effectively respond to most hazards. Others, lacking skills and experience, are dependent on institutional response systems for immediate assistance and long-term recovery. Clearly, both approaches can reduce losses; however, the former strategies often prove more cost effective. How then, can loss reduction be handled with minimal external involvement? Are there not strategies that can reduce the impact of natural hazards and minimize exposure to these risks? Introduction of hazard mitigation strategies can begin to address these concerns in a timely and cost effective manner. Federal institutions in Canada and the United States recognize this potential, and in response, have developed policies and designed programs to limit the impact of hazards. To further collective thinking on hazard mitigation, this paper surveys current legislation and programs in both countries, makes a comparative assessment, and seeks to identify implications for mitigation policy within the framework of the Canadian IDNDR initiative. Perceptions of mitigation are explored to create a common foundation from which to address the nature, direction, and structure of mitigation policy and programs in Canada. Clarity of purpose, awareness of context, and acknowledgment of limitations and conflicts represent crucial aspects in the evolution of mitigation strategies. Through awareness, understanding, and a willingness to engage in dialogue new ways of thinking can be employed to decrease the impact of natural hazards on society and the environment.

Key words: hazard mitigation, federal emergency legislation, mitigation policy, mitigation programs, emergency management, IDNDR.

1. Introduction

The potentially disastrous implications of natural hazards have become of concern to citizens and elected officials in Canada and the United States. The increasing magnitude of losses in recent years has led administrators to reassess coping strategies and ask probing questions. How can long-term loss reduction be most expeditiously realized and sustained? Are there strategies that can reduce the impact of natural hazards and minimize societal exposure? In the face of escalating losses, attention is beginning to shift away from protection and towards an emphasis

* This essay is an abridged version of the original discussion paper funded and published by Emergency Preparedness Canada, an agency of Canada's federal Department of National Defense. The essay does not necessarily reflect the policy point of view of the federal government.

on mitigation. Federal institutions in Canada, and especially the United States, recognize the need for action to minimize losses from future natural disasters. Canadian policy-makers, in the economic interest of the country and in concert with sustainability initiatives, have begun to give disaster mitigation strategies serious consideration. This essay surveys current hazard mitigation legislation and programs in both countries, makes a comparative assessment, and seeks to identify implications for mitigation policy within the framework of the Canadian IDNDR* initiative.

Disasters are not isolated from the social structures within which they occur; rather, they are a social phenomena, the result of human-environment interaction. To achieve tangible results, well-developed partnerships directed towards the achievement of mitigation will be required. To underscore the importance of partnerships, consider the argument that mitigation activities occur, first and foremost, at the local or individual level. Movement towards partnerships is crucial, to share responsibility, minimize risk, and maximize return on the investment of limited resources. Consequently, mitigation becomes a compelling component of federal disaster policy.

Perceptions of mitigation require exploration to create a common foundation from which to address the nature, direction, and structure of mitigation policy and programs. Clarity of purpose, awareness of context, and acknowledgement of limitations and conflicts represent crucial aspects in the evolution of mitigation strategies. Through awareness, understanding, and a willingness to engage in dialogue new ways of thinking can be employed to decrease the impact of natural hazards on society and environment.

During the first half of the International Decade for Natural Disaster Reduction (1990–1995), we have witnessed the severe limitations of our abilities to buffer ourselves from the extreme forces of nature, despite major efforts, a panoply of technologies, and substantial financial expenditures. Without doubt, losses of real property and lives would have been far greater had these efforts not been made. Nonetheless, recent assessments (Bruce, 1994; Burton, 1994; FEMA, 1995a; Showalter *et al.*, 1993; Rossi *et al.*, 1993; IFRC/RCS, 1993) of the magnitude and trend of losses due to natural hazards continue to support the notion of a more hazardous world.

When we focus on a single disaster, whether local or widespread, attention is inexorably drawn to the physical magnitude of the perpetrating forces (i.e. flood height, wind speed, snow depth) and the numerical extent of the impact of these forces (i.e. economic losses; lives lost). Such tendencies are normal, a means of classifying events and comparing impacts among dissimilar forces. Of greater concern though is the tendency of the media to aggrandize disaster which successfully camouflages the underlying causes for the public, and many of the decision-makers involved. In the heat of the moment and under pressure to act,

* The International Decade for Natural Disaster Reduction (IDNDR) is a United Nations sponsored initiative to reduce global losses and human suffering resulting from natural disasters.

answers are given and solutions sought. Time to reflect on contextual linkages is unavailable, probably unnecessary in the fight to save lives and protect property, but ultimately essential to understanding what happened and why.

The inability to grasp the context of events represents one of our greatest barriers to understanding the cause of natural disasters. To view all the factors contributing to a disaster, by cutting across professional and disciplinary boundaries in both time and space, is difficult. Within this essay the concept of mitigation provides a vehicle to explore a contextual approach to the reduction of losses due to natural hazards, and in so doing provide connections to the global discussion on sustainability and quality of life.

2. Mitigation

An appreciation of the various interpretations of the term 'mitigation' is essential to an understanding of initiatives which use this label. Moreover, the attitudes which underlie policy directions will be driven by the definition individuals with decision-making power give to the term. Naturally, problems, errors, and inefficiencies in decision-making arise when people interpret concepts and key words differently. To grapple effectively with the concept of mitigation an appreciation of these differences must be part of the knowledge base of all parties, particularly policy makers and elected officials. Without this semantic version of a 'level playing field', mitigation policies and their implementing programs are destined to evolve slowly, in a confusing disarray, or not at all.

First however, we must understand what is meant by the term 'disaster' and clarify the concept of 'natural disaster'. The importance of clarity in communication* did not escape the denizens of the IDNDR who early in the decade commissioned an internationally agreed glossary of basic terms related to disaster management (United Nations, 1992). Thus 'disaster', as defined in the glossary, means: "a serious disruption of the functioning of society, causing widespread human, material, or environmental losses which exceed the ability of affected society to cope using only its own resources" (United Nations, 1992: 21). Overall this explanation is well crafted, with the exception of the latitude in scale one might assign to 'affected society'. If community, region, or country were substituted the designation of disasters would alter. This distinction may be subtle, but is of importance for those who track levels of disaster occurrence, and especially for people who seek access to disaster financial assistance. Moreover, the level at which disasters are designated may differ between countries as a function of their internal political structures, though when communities can no longer cope, few would deny the existence of a disaster.

* Imagine the difficulty of finding commonality between an array of cultures and languages. For example, the Japanese have no term for mitigation; they want to use 'natural hazard prevention' (Tsai, 1996). Definitions are important and cannot be taken for granted. The inclusion of definitions in most pieces of legislation testifies to this fact.

To advance discussion we need to rethink, in a fundamental way, what is meant by a 'natural disaster' for the way we think about natural disasters will influence where we look for solutions, and the shape and character of the means we use to attain those solutions. Over time attribution of natural disasters has shifted from supernatural or mystical forces, to nature (physical forces in natural systems), and with some reluctance, to humans who have made changes to natural systems. In this transition of thinking we have added to the list of causation, but never fully abandoned the earlier sources. Quarantelli (1989) notes the inherent danger in this approach when he observes that "the distinction often drawn, between so-called Acts of God (or Nature) and Acts of Men and Women is both a useless and false one.... There also lurks in the distinction a supposition that one kind of disaster is more directly controllable than other ones" (1989: 2).

Moreover, a natural disaster, in a pure sense does not exist; rather there is the interaction of changes in physical systems with existent social conditions. Disasters are therefore more accurately seen as social phenomena. The implications of rethinking 'natural' disasters as having social causation is, in Quarantelli's (1989) view, fourfold:

(1) mitigation of disasters must stress social rather than physical approaches;
(2) these approaches must place emphasis on pro-active rather than reactive actions;
(3) such actions need to focus on internal flaws in society rather than external forces; and,
(4) reduction of vulnerability to disasters must be integrated as part of ongoing policies and programs of societal development.

If then, we accept that full prevention is unattainable, our rethinking of disasters leads us towards a policy of long-term loss reduction – *mitigation*.

Understanding how 'mitigation' is interpreted thus becomes of utmost importance. The UN Department of Humanitarian Affairs defines mitigation as: "[m]easures taken in advance of a disaster aimed at decreasing or eliminating its impact on society and environment" (1992: 41). The conciseness of the IDNDR definition is attractive, though the reference to 'eliminating' is disconcerting in light of persuasive arguments (Hewson, 1996; Vollard, 1996; Quarantelli, 1989) that the impact of disasters cannot be fully prevented, but only minimized.

In general usage, efforts to 'mitigate' disasters are considered positive actions to reduce the severity, violence, cruelty, or pain of prevailing or future conditions. Acknowledgment of widely accepted interpretations does not restrain individuals and professional organizations from developing their own meanings for important terms. In highlighting these differences there resides no inherent attempt to provide a definitive judgment, but rather a need to point out the diversity of views involved and the resultant problems this can cause for communication, policy development and decision-making. Awareness is the first step towards understanding; thus identifying basic differences can lead to productive discussion within and between organizational units, and ultimately wiser decisions. Such differences underline the importance of definitions and interpretation in the formulation of public policy.

In Canada perceptions of mitigation are diffuse and individual, in the absence of any formal policy statement or organizational delivery structure at the federal level.* Some officials see mitigation as primarily post-disaster physical actions taken to raise dykes or reinforce structures to withstand wind and vibration (Peters, 1996; Shea, 1996), while others take a nonstructural perspective. The need to involve all constituencies from the individual to the corporate, and utilize both structural and non-structural loss reduction mechanisms was mentioned by deBeaupré (1996) and Simard (1996). Regardless of perspective, when mitigative activities do occur in Canada they tend to be disaster-specific and at the post-recovery stage of the emergency management cycle.

The confusing usage of the term 'mitigation' in the Canadian hazards community was raised by Burton (1996) who felt specific meanings of mitigation would arise from different contexts. If, by context Burton is referring to variation in the beliefs and values held by decision-makers, then certainly different meanings are likely to arise. Actions to achieve loss reduction, however, will vary for specific hazards such as tornadoes or earthquakes, and differences of interpretation can clearly affect the nature of these actions, and, thus, the resultant outcomes. The lack of widespread clarity on the relationship between mitigation and other stages resulting from human-hazard interaction (i.e. preparation, response, recovery) contributes to the confusion Burton (1996) mentions. Peters (1996), however, makes a clear distinction, locating mitigation as a post-recovery activity in the emergency management cycle, i.e. occurring at some stage after an initial disaster. Where one locates mitigation on an emergency management cycle is a moot point, as direct mitigative actions are not currently a primary focus of Emergency Preparedness Canada's mandate.

In contrast, the Federal Emergency Management Agency (FEMA) considers mitigation a more pervasive activity and thus a cornerstone of its loss reduction strategy (FEMA, 1994a; 1995b), though some would debate the effectiveness of hazard mitigation in light of the overly generous federal disaster assistance policies (Platt 1996). Most discussants paraphrased the approved agency definition of hazard mitigation which is "sustained action taken to reduce or eliminate long-term risk to people and their property from hazards and their effects" (FEMA 1995a: iii). As with the IDNDR definition the use of 'eliminate' is disconcerting and indicative of a control-oriented tendency. The need for clarity in defining mitigation was emphasized by Moore (1996) as States, in their own interest, attempt to extend the meaning to include capital purchases, such as warning sirens, as part of costs eligible for federal mitigation support. This raises the much discussed distinction

* While this paper will limit its scope of attention to the federal level a similar lack of policy and implementing legislation is evident in provincial jurisdictions (Newton, 1995: 65). This is not meant to imply that mitigation style activities do not occur, but rather they are not formally mandated and thus occur as a result of individual action.

between mitigation, preparedness, response and recovery, as well as terms such as rehabilitation and restoration (Rubin, 1985). In organizational structures with delegated responsibilities and financial allocations, these distinctions are important to politicians, program administrators, and the individuals who have suffered losses. Semantics aside, a case can be made within the scope of FEMA's definition, for mitigation to be a component of the other terms, provided effort or expenditure results in the *long-term reduction of risk*. Thus, temporary sand bags would not be considered mitigation, while floodproofing would, regardless of when these actions occurred. From a program implementation perspective Shea (1996) felt mitigation involved actions that helped strengthen the built environment (control-orientation) through building stronger infrastructure in a smarter manner. However, for many administrators in the United States the term 'mitigation' has only recently entered their daily lexicon, making application experimental, and problematic, as they adjust to new attitudes and decision-making criteria. In view of the high potential for differing interpretations of the term 'mitigation', "it is important to note that in the United States ... it refers only to measures taken independent from an emergency situation or actual disaster" (National Research Council, 1994: 21).

To summarize the foregoing insights about the meaning of 'mitigation' one is drawn to a simple, straight-forward definition, which touches the heart of the matter – mitigation is about risk reduction and a resultant decrease is future losses. With due observance to the timing of mitigative activity mentioned above, mitigation is pre-meditated actions leading to a reduced risk of a loss occurring. The term pre-meditated is intended to distinguish mitigative actions from activities during the pre-disaster preparedness period and the recovery period where losses can also be reduced. As well, the emphasis here is on the 'reduction of losses' not the elimination of risk which has tended to be the focus of institutional agencies. The fact is, most risks cannot be eliminated, and recent disasters, particularly due to flooding, demonstrate that the residual risk is greater than anticipated.

Fundamentally though, disasters still trigger the flow of federal mitigation funding, and thus FEMA has placed an emphasis on immediate post-disaster activities when availability of funding, political will, and individual incentive are all high. Myers (1996) sees this as a pragmatic but unfortunate approach as, during non-disaster periods, mitigation is less likely to be included as part of ordinary day-to-day decisions and policy development. Upgrading buildings during post-disaster reconstruction and repair are likely to receive greater attention than improvements to building codes and modifications to land use practices. Whether used in Canada or the United States 'mitigation' will require ongoing clarification and long-term usage to be widely accepted and understood.

3. Recent Trends in Disaster Mitigation

The occurrence of major disasters such as the Kobé and Northridge earthquakes, hurricane Andrew, the mid-west U.S. floods of 1993 and 1995, and the Saguenay flood of 1996 have raised difficult questions about the perceived safety of human settlements. Are current actions sufficient? Are they the most appropriate actions? What might we do differently? Against a backdrop of escalating financial losses (Bruce, 1994; FEMA, 1995a) industrialized countries are beginning to reconsider these questions and assess options of how they might respond. Such contemplation is not new, but now it is being intensified as a result of two factors. First, the International Decade for Natural Disaster Reduction has raised political awareness of natural disasters and their consequences on a global scale (Lechat, 1990). In their own context each political body is asking – How are we doing? Many would not like the answers. Second, the previously unthinkable billion dollar losses have become realities, not just in overall losses, but in insured losses and government payouts. Natural disasters at this scale are no longer regional issues; they have become big ticket items of national significance, which existing control systems are unable to handle.

3.1. THE CANADIAN SITUATION

Immunity from extreme natural phenomena which result in losses of lives and property is unrealistic for any country, especially one with the geographic scope of Canada. However, the uneven distribution of the Canadian population concentrates the potential for significant losses, along with public pressure and political sensitivity, to a small portion of the country. As well, a climate of extremes has conditioned the populace to adapt to a variety of severe weather conditions more readily than jurisdictions with a moderate climate.*

Overall, Canadians experience less risk due to natural disasters than Americans or islanders in the Caribbean. Nonetheless, disasters do occur throughout the country, from tornadoes in the west and southern Ontario, to flooding on numerous rivers, and hail in the lee of the Rocky Mountains.** The potential implications of such hazards cannot be ignored by administrators and politicians as each has responsibilities for securing the safety of life and property. However, with neither the humanitarian pull of thousands of people in need and unable to cope, nor the financial push of staggering losses, impetus at all levels of government has been moderated to extension of existing services, rather than expansion. A critical mass of loss on the scale of an Andrew or Kobé has yet to materialize in Canada. The inherent contradiction in this condition is that Canada is fortunate and yet becoming increasingly vulnerable at the same time. A brief review of trends in emergency policy and programs in Canada is warranted.

* Heavy snows during January 1996 in the American northwest caused severe disruption to major centres such as Philadelphia, Baltimore, and Washington, D.C. Canadian cities routinely cope with such conditions.

** The largest single insured loss to date in Canada was the 7 September 1991 hailstorm which hit Calgary. The insured payout was over $360 million (Ross, 1994: 4–5). Losses were high in the 1996 Saguenay, Quebec flood (bills are still being calculated at publication date). By comparison the insured payout for Kobé was $3 billion.

The initial impetus to develop emergency planning and response activities in Canada resulted from the 1948 formation of a civil defence organization as part of Canadian participation in NATO (EPC, 1996a,b). Until the mid-1960s Canadian civil defence agencies focused their attention on the threat of nuclear explosions (Government of Canada 1989a). By 1966, the need for a coordinated federal response to peacetime disaster was recognized and assigned to the Emergency Measures Organization. Within ten years the focus had shifted to peacetime emergencies and further wartime actions put on hold. Through these efforts, disaster planning for peacetime emergencies received increased attention leading to national legislation, mutual aid agreements and provincial response agencies throughout Canada.

Federal emergency policy was designed to address a spectrum of disasters while vesting primary responsibility in local, provincial and territorial governments. The federal government objectives were: (1) to provide leadership for improved emergency preparedness in general; (2) to develop a credible national capability to meet emergencies of all types; (3) to realize reasonably uniform standards of emergency services across the country; and, (4) to be sensitive to humanitarian concerns (Government of Canada 1989b). The replacement of the *War Measures Act* by a new comprehensive *Emergencies Act* (1988) shifted the emphasis further towards confronting the increasing incidence of natural disasters in Canada. The subsequent enactment of the *Emergency Preparedness Act* (1988) provided the necessary legislative framework for current emergency preparedness activities throughout Canada (see Table II) In its civil application the Act, while mentioning activities some might interpret as mitigative, does not address mitigation directly, choosing rather to develop the preparatory character of existing federal initiatives. In contrast, Canada's contribution to the IDNDR mid-term review in Yokohama (Canadian National Committee for IDNDR, 1994) includes a major section on mitigation activities. Review of this section reveals a broad use of the term 'mitigation' to include response, preparedness and planning activities, many with short time horizons and a focus on providing information that might encourage loss reduction actions. With the exception of the discussion of drought, the initiatives addressed tend not to provide long-term sustainable solutions which can be effectively implemented by local governments and individuals.

At present, an integrated national loss reduction strategy remains elusive in the Canadian hazards community though few, if any, would deny the necessity of mitigative actions. The evolution of Canadian federal action in the natural hazards field has historically taken focuses which, while including activities with mitigative components, have not specifically targeted mitigation in a comprehensive manner.

3.2. MITIGATION EXPERIENCE IN THE UNITED STATES

Some of the most aesthetically and economically attractive locations in the United States are also the most risky in terms of vulnerability to natural disasters. Coastal

Table I. Federal natural hazards legislation

Canada:
Canada Water Act 1970
Emergencies Act 1988
Emergency Preparedness Act 1988
United States:
Coastal Zone Management Act 1972, as amended
The National Flood Insurance Act 1968, as amended
The Flood Disaster Protection Act 1973, as amended
The Federal Disaster Relief Act 1974
The Federal Fire Prevention and Control Act 1974
The National Earthquake Hazards Reduction Act 1977
The Robert T. Stafford Disaster Relief and Emergency Assistance Act 1988, as amended
The National Flood Insurance Reform Act 1994
Executive Order 11988, Floodplain Management
Executive Order 12941, Seismic Safety of Federally Owned or Leased Building
Executive Order 11990, Protection of Wetlands
Executive Order 12699, Seismic Safety of Federal and Federally Assisted or Regulated New Building Construction
Joint Agreements
Canada/United States Agreement on Emergency Planning 1986

Sources: Tubbesing (1989), Fowler (1993), FEMA (1995b).

Table II. Federal natural hazards programs

Canada:
Canadian Emergency Preparedness College Programs
Disaster Financial Assistance Arrangements
Joint Emergency Preparedness Program
National Earthquake Support Plan
National Emergency Arrangements
National Flood Damage Reduction Program
Vital Points Program
United States
FEMA Mitigation Information Services System
Hazard Mitigation Education and Technology Transfer Programs
Hazard Mitigation Grant Program
Interagency Hurricane Evacuation Program
Interagency Post-Disaster Hazard Mitigation Teams
National Earthquake Hazards Reduction Program
The National Hurricane Program
USGS Hazard Warning Program

Sources: Tubbesing (1989), Fowler (1993), FEMA (1995b).

areas are subject to hurricanes in the east and earthquakes in the west, while riverside settlements are exposed to flooding. Americans know these risks exist, due to historical losses.* The magnitude of losses during the last six years was staggering and unprecedented in United States' history, reaching the $20–30 billion range for a single disaster. Increases in population and investment in established metropolitan areas are partly responsible for this sharp upward trend.** Clearly settled areas are more vulnerable than realized, despite years of effort to protect life and property. It is important to note though, that without preparatory and protective efforts many more lives would have been lost and property damaged.

The trend in America of reduced annual fatalities and rapidly increasing property losses due to natural disasters has been well documented (Mileti *et al.*, 1995; Rossi, *et al.*, 1995; Witt and Rubin, 1995; Ross, 1994; Showalter *et al.*, 1993). Can we then say that the United States is becoming a safer place to live? Probably, yes. And yet the risk due to natural hazards is geographically skewed making dense population centres which coincide with high hazard potential (Florida and southern California, for example) riskier places to live. The Clinton Administration, with roots in the south, was well aware of this dilemma when it took office in 1992. A new course of action, specifically designed to mitigate future losses, was needed to complement and extend existing federal programs (see Table II).

Federal efforts to reduce damage due to natural hazards began to evolve in this direction when James L. Witt was confirmed as the new Director of FEMA. A clear indication of change came on 29 November 1993 when the profile of mitigation activities was raised significantly with the formation of a new Mitigation Directorate. The initial program structure of the Directorate came from programs resident in the former State and Local Programs and Support Directorate (SLPS), and the Federal Insurance Administration (FIA) (FEMA 1994a). The amalgamation of disparate programs with a strong mitigation content was to be the hallmark of the new Directorate's evolution. To shift, however, from the more common programmatic thrust of governmental service delivery structures to a functional orientation more suited to the achievement of mitigation objectives would prove an immense undertaking. Today, this policy direction continues to be implemented and debated.‡

---

* Hurricanes: Hugo 1989; Andrew 1992; Iniki 1992. Earthquakes: Loma Preita 1989; Northridge 1994. Floods: Mid-West 1993 & 1995; Northeast 1996. Wildfires: Santa Barbara 1990; Oakland/Berkeley 1991; Southern California 1993.

** Other factors fueling the upward trend in losses include: insufficiency of current infrastructure; poor planning; post-loss gouging; fraud; the fragility of structures; lack of attention to proper maintenance; and, the ability of insurers to respond due to bankruptcy (Ross 1996).

‡ Platt (1996) notes that federal disaster policy critiques have come from the National Academy of Public Administration, the National Performance Review, the General Accounting Office, the Congressional Research Service, the Interagency Floodplain Management Task Force, the National Earthquake Hazard Reduction Program (NEHRP) Advisory Committee, the 'Galloway Committee' that examined the Midwest floods of 1993, the House and Senate Task Forces on Disaster Assistance, and the FEMA Inspector General.

FEMA's commitment to mitigation moved ahead rapidly, driven alternately by the unfortunate occurrence of large-scale disasters (see footnote #8) and the vision of the new Director. Within 24 months of its inception, the Mitigation Directorate produced the *National Mitigation Strategy – Partnerships for Building Safer Communities* (FEMA 1995a). At the first National Mitigation Conference, attended by over 850 federal, state and local emergency managers, Witt reiterated his commitment: "Mitigation is a priority for FEMA. Mitigation must go from a little used word after a disaster strikes – to a household word 365 days a year. But to do that, I need your continued support, and your commitment to work in partnership with the federal government" (Witt, 1995). The reference to 'partnerships' underscores the FEMA strategy, indicating that a greater sharing of risks and responsibilities would be crucial to the success of mitigation initiatives. Moreover, in promoting this concept of sharing, the Directorate would be attempting to bring about a fundamental change in the perception of hazard risk by the general public (FEMA, 1995a). The prevailing perception at state and local levels "that the costs of disasters will be predominantly absorbed by the nation" (Platt, 1996) would have to change. In light of the short term memory of many citizens who continue to settle hazardous sites, this change would become pivotal to the strategy. To alter public perception would demand radical thinking by generally conservative institutions. Overcoming inertia represents a significant challenge.

Changes to public perception were not the only major adjustments in thinking needed. The Directorate itself, after months of deliberation had decided to embark on a functional structure and abandon the historical programmatic arrangements familiar to clients and Congress alike. To deliver the new message clearly and coherently in times of stress, and during the all too infrequent inter-disaster periods, Directorate staff needed to have their thinking process rejigged. Cutting across program lines became the norm for Mitigation Directorate managers and staff. Convincing others became an ongoing challenge, and the structure of the legislative authorities did not help.

In establishing a Mitigation Directorate and choosing to operate from a functional foundation, Witt selected a theoretically sound approach whose implementation requires delicate manoeuvring through an operational and legislative minefield. No single Act provides the basis for this undertaking. Rather, the Directorate must achieve its objectives based on the authorities provided by a polyglot of existing legislation. For daily operations, the Mitigation Directorate draws authority from a web of legislation including the Flood Disaster Protection Act (1973), the National Earthquake Hazards Reduction Act (1977), annual appropriations for the National Hurricane Program, and the Federal Fire Prevention and Control Act (1974), together with numerous Executive Orders and other authorities (see Table I) put in place since the first permanent disaster relief legislation was enacted by Congress in 1950 (FEMA, 1995a). FEMA is well aware of the limitations of such a fragmented legislative basis for the implementation of the National Mitigation Strategy. One important breakthrough, possibly a harbinger of things to come, was the National

Flood Insurance Reform Act (1994), which amended the Stafford Act to broaden the base of federal disaster assistance on which mitigation funding is calculated, *and* enhanced the amount dedicated to mitigation from 10 to 15 percent of total federal expenditure on a disaster. The implication of these changes quickly became apparent in the aftermath of the Northridge earthquake when in excess of $700 million was made available for mitigation activities in devastated counties, albeit on a cost-shared basis.*

Though prepared in response to proposed legislation for a national all-hazards insurance program, the policy paper *Natural Disaster Insurance and Related Issues* (Witt and Rubin, 1995) incorporated a substantive mitigation emphasis which could provide a basis for future discussions with Congress on legislative reform. Reform of federal disaster policy is being led by Representatives Emerson and Mineta (1995). Together with a broad coalition of private sector interests they are currently working on further legislative reform that would dramatically change federal disaster policy, including the establishment of a new, privately-funded entity, known as the National Disaster Insurance Corporation (NDIC). The Emerson-Mineta proposal, through a new Natural Disaster Protection Partnership Act, seeks to address the escalation of disaster relief since 1989 and the weaknesses in the available means of response raised by Witt and Rubin. For example, as individual perception guides decision-making, current nonemergency federal practices can be disincentives to mitigative action. Witt and Rubin note that: "[t]he expectation of Federal assistance, while reassuring to potential disaster victims, can also be a factor in consumers' decisions about purchasing insurance, locating prudently, and building protectively" (1995: 3).

Similar reasoning may influence the maintenance and renovation of public infrastructure in hazardous locations throughout the United States. Platt (1996) concludes that "federal benevolence discourages responsible hazard mitigation among non-federal interests, thus contributing to the potential for greater losses in future disasters". Clearly, this was not the intention of federal disaster assistance and, thus, the Administration's policy paper discusses proposed actions "in five major areas – loss prevention, compensation for losses, supplementary federal protection for catastrophic losses, federal assistance after a disaster, and program evaluation" (Witt and Rubin, 1995: 5). The initiatives proposed are thoughtful, aimed at helping mitigate the devastation caused by disasters in a targeted manner that reflects an appropriate sharing of responsibility. Unfortunately, budget issues appear to have become the driving force behind decision-making at FEMA, leaving the future of reforms to hazard legislation uncertain though still an important policy area to congressional representatives. On a more promising note, the recent elevation of the position of Director of FEMA to the cabinet level may allow for a more rapid advancement of a mitigation-based loss reduction agenda.

* The normal federal/state split is 75/25 though this can be altered up to 100/0 by Presidential order.

4. An Assessment of Current Mitigation Practices in North America

The value of comparisons lies in the expectation that new ideas or problem solving approaches that arise within one jurisdiction or locale, can be profitably applied to other areas of jurisdiction or operation. The challenge to realizing this benefit is to surmount the dissimilar natures of the jurisdictions. Too often, in an attempt to create a more equitable basis for assessment, differences become minimized rather than valued, leading to skewed outcomes. To avoid this tendency, federal activities which mitigate the impact of natural hazards in Canada and the United States will be considered collectively at three levels of detail, namely: (1) the large scale influential forces which guide governmental decision-making in the hazards sector; (2) the policy and legislative positions moulded by these influential forces; and (3) programs that legislation has spawned to reduce long-term risk to human life and property.

4.1. INFLUENTIAL FORCES

Historically, in both Canada and the United States federal actions associated with natural hazards have focused on saving lives, and helping people get back on their feet after catastrophes beyond the immediate coping capability of provinces or states. Notwithstanding the presence of some mitigative activities (e.g. reconstruction to a higher standard) most actions were oriented to rapid recovery. In these times of crisis, human need and legislative authority drove the process. The forces influencing mitigation evolved in a somewhat different, more diversified manner, the magnitude of which varies markedly in the jurisdictions under consideration.

First, large scale natural phenomena* such as hurricanes, tornadoes, earthquakes, wildfire and floods seem to be occurring more frequently in recent years, as discussed previously, but are not distributed equally throughout North America. Frequency by itself might be only mildly concerning, if the prime locations of occurrence were not also areas of substantial growing populations. In the United States the frequency at which extreme natural conditions and large populations intersect far exceeds the occurrence in Canada. Hurricanes tend to batter communities along the Gulf of Mexico or the eastern seaboard and earthquake and volcano activity is highest in Southern California and the United States' Rockies. Simply put, more people in the United States are exposed to a variety of extreme natural forces than in Canada.

Second, the magnitude of resultant economic losses and the increasing exposure of capital investment, has transformed what were at one time losses of regional significance into situations of national concern. Again, such conditions appear more

* We must however, acknowledge the implications of human intervention which can influence the occurrence and magnitude of extreme natural episodes (Hewitt, 1983). For example, deforestation, use of impervious surfaces, and channelization can increase and speed-up storm run-off resulting in higher water levels and potentially disastrous episodes of flooding. Similarly, unnatural build-up of forest refuse creates the potential for conflagrations.

prevalent in the United States than in Canada. Investment and building continues unabated adjacent to the coastline from Boston to Miami, in river valleys across the United States, and in the already heavily built counties of California. Canadian capital investment is small by comparison, and generally dispersed with greatest concentrations in Montreal, the Greater Toronto Area, and the lower mainland of British Columbia. Of these areas, only the lower mainland has a high potential for natural disaster due to earthquake, riverine flooding, tsunami, or a combination. Economic losses are naturally of concern to Canadian administrators who support the need to minimize them, though not to the extent that the recent multi-billion dollar disasters are causing United States agencies such as FEMA to rethink their mandate.

Third, and directly attributable to the previous two points, is the manner in which political systems not only address problems, but rank them. New policy directions require political will, which itself needs both a strong public push and a significant political currency. Virtually all proposals for federal expenditure of new money will have to pass this two-tiered test – wide-spread public support (relative to other possible expenditures) and substantive political value (generally related to re-election). Politics has always been an influencing factor whatever the area of focus; the mitigation of natural hazards is no different.

Fourth, observation of and research on disasters in Canada (Newton, 1995), and more recently in the United States (e.g. the Northeast Storm of '96 and the resultant flooding), tend to indicate a more substantial federal involvement in the United States than in Canada. While there are similarities between how the respective federal governments deal with provinces or states in the event of a disaster, there appears to be a greater federal presence, both political and administrative, in the United States. Disasters seem to have a high political currency in the United States, whereas the federal profile is relatively low in Canada. Likely this is a result of the respective balance of power between federal and provincial or state governments and fundamental differences in how our respective countries are governed. The resultant level of involvement represents one of the most important factors influencing the current state of federal mitigation policy and legislation in both counties.

4.2. MITIGATION POLICY AND LEGISLATION

How one determines whether legislation does or does not contribute to the mitigation of natural disasters appears to be driven less by research than by differing interpretations and understanding of what 'mitigation' is. Alternative views on this subject were discussed earlier. If one compares the Canadian Mid-Term Report (Canadian National Committee for IDNDR, 1994) and the United States National Mitigation Strategy (FEMA, 1995a) 'mitigation' is interpreted somewhat differently, with the Canadian report (see Chapter 3) casting a very broad net. With such an apparent variation of interpretation, direct comparisons are not productive. One can only state that the existing Canadian legislation, having evolved during the 1980s (Government of Canada, 1988a; 1988b), is silent on the subject of mitigation while the United States has only recently amended the Stafford Act (Government of the United States, 1988) to enhance the potential for mitigation.

In light of the limited attention to mitigation within legislative authorities, our attention might better be directed to general characteristics, more specifically the structure, flexibility, and development of hazard-related legislation. This approach may begin to reveal the intention of the administrators with greater clarity and, in so doing, provide further insights into how specific mitigation activities may or may not fit informal and formal loss reduction strategies.

Structure. In Canada the *Emergency Preparedness Act* functions as general enabling legislation and an administrative instrument to identify the responsibilities of federal agencies (EPC, 1996a: 6). By comparison the United States Stafford Act sets out specific activities to be performed by federal agencies together with guidelines and limitations. In structure, the Canadian legislation provides a framework of legislative authority within which actions may be taken in support of provincial/territorial programs and activities. The feeling imparted is one of a basis upon which responsible agencies have latitude to make decisions related to the prevailing set of circumstances, and a mechanism to permit the flow of emergency financial assistance. In the United States, despite legislation that provides more definitive operational direction to states and local governments, disaster assistance payments come primarily from federal subsidies, grants, programs and other benefits (Platt, 1996). In this case, the legislation is action-oriented, with sections on 'Availability of Materials', 'Hazard Mitigation' (404) and 'Repair, Restoration, and Replacement of Damaged Facilities' (406). The contrasting approach of these documents is reflective of the differing federal involvement in regional emergencies. Federal officials in Canada facilitate response and support lead agencies, whereas FEMA has substantive, direct involvement with local governments and individual citizens.

Flexibility. Good legislation is crafted with unambiguous terms which convey the intent of the authors with clarity and internal coherence. The message and application are well stated, and yet there remains an acceptable degree of interpretation – the flexibility needed to address the variety of conditions to which the legislation will apply. In the realm of natural hazards this need is most apparent. Canadian federal emergency legislation leaves the impression of flexibility in interpretation, with the assessment of what is and is not acceptable expenditure negotiated later by provincial and federal administrators using program guidelines. Moreover, the initial financial responsibility of provinces/territories places greater decision making power for disaster response in their hands. Legislative guidance for mitigation activities is ultimately flexible in its absence and, thus, the federal government appears largely dependant on individual and local actions to reduce future losses.

United States' federal legislation is more definitive, especially where mitigation is concerned (Government of the United States, 1988: 14–19). Flexibility through negotiation is less evident; however, the President has some latitude in designating disaster areas on a county by county basis, and modifying the federal/state share of disaster expenditures. On individual claims however, interpretive latitude continues to exist in the determination of eligible expenses under disaster relief programs in Canada and the United States. Such flexibility, though, is not a function of federal legislation but the delivery mechanisms of specific programs.

Development. The magnitude of recent losses and frequency of natural disasters has fostered the development of federal mitigation legislation in the United States. The specific inclusion of mitigation in the Stafford Act, the establishment of a Mitigation Directorate in FEMA, and the announcement of a National Mitigation Strategy demonstrate the commitment of the Administration to the active development and enhancement of mitigation to avert future losses due to natural disasters. However, the elevated profile of mitigation initiatives in the United States cannot be taken as indicative of unconditional support, as historical control-oriented programs are still active and heavily funded. Also, there has been scepticism as to whether the concepts of persuasion and partnerships outlined in the National Mitigation Strategy are materialising, given the continued high level of federal disaster assistance in the United States, (Platt, 1996; Witt and Rubin, 1995). Nonetheless, the Administration appears serious about promoting mitigation as a cost-effective approach to safer communities.

Such a focused development of mitigation legislation, policies and programs has yet to evolve in Canada. To date, development of mitigation initiatives has been constrained by the apparent success of historical disaster reduction practices, as measured by the level of federal payouts. Neville (1996) indicates total federal disaster assistance since inception to be in the range of $262 million. Consequently, development of mitigation measures to reduce disaster losses has lacked the pull of public demand and the push of escalating federal expenditures.

4.3. IMPLEMENTING PROGRAMS

If one interprets 'mitigation' to be specific actions which directly reduce the long-term risk of loss due to natural hazards, then one is left to conclude that implementing programs have yet to be developed in Canada and have only recently been started in the United States. For example, Section 404 of the Stafford Act allows for the acquisition or relocation of private property determined to be at risk. In accepting a settlement under this program the government requires that the property "be dedicated and maintained in perpetuity for use that is compatible with open space, recreational, or wetlands management practices" (Government of the United States, 1988: 15). More frequently, given the high cost of acquisition, federal agencies provide information and education in the hope that people will take responsible action, when they locate and construct buildings.

Brochures, fact sheets, case studies, and emergency plans are common tools in Canada and the United States to promote awareness and encourage effective action when faced with a hazard. Some might consider these to be mitigative measures; others would not. Similarly, communication technology, warning systems, command vehicles, and specialized clothing and equipment are primarily related to direct response and immediate recovery, rather than long-term loss reduction. Movement towards a strategy based on mitigation is not easily achieved. Inter-jurisdictional debate over implementation of loss reduction strategies, influenced by historical turf and existing operational mandates, does have the effect of limiting imprudent initiatives. But, on the negative side, it also limits change in historical patterns of hazard management.

5. Implications for Disaster Mitigation in Canada

The current American mitigation strategy provides a timely thrust for consideration by federal agencies with delegated responsibilities for protecting the health and safety of Canadian society. For most agencies, embracing mitigation will involve a shift in thinking. Current activities would require modification from primarily preparatory actions and response to disasters as they unfold, to a more anticipatory condition with an emphasis on improved local resilience, adaptation strategies, and loss avoidance, rather than control and tactical management. The establishment of a mitigative approach in Canadian agencies is therefore uncertain, subject to change, and at present lacking a clearly articulated holistic vision. Under these conditions, to adopt the United States' approach to mitigation unquestioningly would be imprudent, and possibly inappropriate.

Without the strong pull of escalating economic losses due to natural disasters prevalent in the United States, the impetus for developing a national mitigation strategy in Canada is lessened. Should, however, Canada experience a single event resulting in a multi-billion dollar loss, complete with the trailing fallout for the insurance and banking sectors,*mitigative measures would be demanded by the public, as well as private and public agencies*. Such an event is inevitable; it is just a matter of time. An earthquake in the lower mainland of British Columbia represents the most talked-about candidate, with loss estimates in the $14–32 billion range; though a hurricane tracking inland through Ontario's Golden Horseshoe, or the Montreal region also would create devastating losses. While the magnitude of such losses is a compelling reason for action, human tendency is to prepare for emergencies only to the extent warranted by experience, and to respond when a disaster takes place. The success of this approach in Canada has also been its greatest weakness; every success reinforces faith in the preparatory measures taken and in the existing response system, while increasing vulnerability to large-scale low probability events. Part of the challenge in moving towards the incorporation of mitigation in a loss reduction strategy, is to overcome the complacency that success breeds.

Despite ambivalence about their identity, Canadians as a people are different from Americans. Our approaches to problem-solving differ, even for the same problem. However, the American system of government concentrates power, leading to greater direct federal involvement in the formation of legislation and the disbursement of funds, than is seen in Canada. A hallmark of the Canadian process is the highly consultative nature of policy development and the extent of powers delegated to the provinces and territories. Extensive consultation, while less prominent in recent years as fiscal realities hold sway, remains an expectation within all Canadian jurisdictions.

No longer is government seen to be omnipotent, responsible for all ills, and yet, all at the same time, a financial magician. Increasingly, the form and nature of governance is shifting toward an even greater dependency on partnerships. Not just with other governments but with private corporations, non-government organizations, unions, communities, families, and individuals. With partnership comes a sharing of responsibility, a sharing of risk, and hopefully a greater resilience to what lies ahead. Discussion of a mitigation strategy must evolve within this context and reflect the fundamental shifts in attitude and public perception that are occurring. Future disasters will not be isolated from the social structure within which they occur; rather, as Quarantelli (1989) has noted, they will be seen increasingly as social phenomena. Numerous different constituencies (e.g. financial, insurance, corporate, community leaders, potential victims, etc.) will have to be an integral part of disaster mitigation initiatives. Certainly, the federal government still has a lead role and will be expected to accept overarching responsibility for initiating and guiding policy developments and implementing actions at the national level. To achieve tangible results, well-developed partnerships directed towards the achievement of mitigation will be required. The growing complexity and diversity of Canadian society make success unlikely with a single agent.

To underscore the importance of partnerships, consider the argument that all mitigation is *local*, meaning the action needed to implement a loss reduction idea will occur at the local or individual level, where the need is evident, and the benefit will be felt first. Historically, people on the frontier were almost totally responsible for themselves; they acted in their own best interest, without external assistance. Such individual strength is rare. Today, partnerships will be essential to the realization of successful mitigation activities. Working together, building those partnerships represents one of the underlying requirements of a mitigation strategy.

Mitigation of human and property losses due to natural hazards will require modified thinking, collectively and individually. Articulation of the objectives, which revolve around the development and enhancement of local resilience and adaptive capability, are well within our normal frame of thinking and policy development. However, given declining federal program spending, innovative thinking will be required to seek low-cost/no-cost means of implementing loss reduction strategies. Thus, a move towards partnerships, to spread responsibility, minimize risk, and maximize return on the investment of limited resources becomes a compelling component of a mitigation strategy. The challenge of building resilience to natural hazards in the general populace and in the quality of the built environment, rests to quite some degree on modified thinking, or more specifically changes in perception of human-environment interaction. Only through shifts in perception (considered a prerequisite for adaptation) can a state of mind exist where mitigative actions can be effectively taken and a reduction in losses realized.

As background to this evolutionary process of survival, Mileti *et al.* (1995) explore the integration of natural hazards and sustainability for communities at risk in the United States. The investigation is sufficiently conceptual as to be applicable to thinking on this subject in Canada. In essence, the thesis states that human sustainability cannot be achieved without attention to natural hazards and the development of a symbiotic relationship. Reaching for this condition, "... calls for a different perspective and cultural changes to support it. In order for the concept to become part of daily life, shifts would have to occur in basic human values, practices, and approaches that guide society-environment interactions" (Mileti *et al.*, 1995: 124). Seeking basic change in society is highly problematic, especially through governmental policy, yet crucial to the long-term safety of Canadians and the prudent fiscal management of limited resources. Attention to this issue is not optional. Either the commitment is made voluntarily, or a major disaster with unacceptable human and financial losses (i.e. Canada's first multi-billion dollar loss) will underline the necessity of changes to our basic thinking about natural hazards. Unfortunately, it is only a matter of time. The question is not whether there will be losses, but what their magnitude will be.

To address this problem, the temptation to marry private entrepreneurship with public accountability is enticing, though fraught with discontinuities and potentially destructive overlaps between guardian and commercial syndromes (Michleburg, 1996; Jacobs, 1992). As Jacobs (1992) astutely observes, conflict occurs when precepts appropriate to one syndrome are adopted by the other. Public bodies need a cautious approach to private practices, provided perhaps, by the establishment of sensitive partnerships with practical objectives and a clarity of purpose characteristic of market leaders.* Examples of efforts in this direction include the various federal and provincial round tables on environment and economy, and mechanisms to set and implement national and international standards (e.g. building codes, CSA designations, ISO9000 – quality production standards, and ISO14000 – quality management standards). Mitigation initiatives would be well advised to consider similar co-operative approaches.

---

* Attention to recent business literature can be more profitable than publications that attempt to redesign government through utopian privatization. Thoughtful interpretation within the governmental context is required by public leaders rather than amputation of services in the guise of efficiency and fiscal responsibility. One would be advised to balance the forthright guidance of Treacy and Wiersema (1995) in *The Discipline of Market Leaders* and Whiteley (1991) in *The Customer Driven Company* against the pitfalls Jacobs (1992) presents in *Systems of Survival.*

Towards the end of the United States National Mitigation Strategy (FEMA, 1995a: 23) and the Administration Policy Paper (Witt and Rubin, 1995: 17), acknowledgement of the importance of evaluation is found as a fitting rounding out of each document. The rationale Witt and Rubin (1995) supply in support of mandated evaluations speaks strongly in concept to any mitigation initiative:

> Because the effectiveness of these [mitigation] initiatives, is in part dependent on complex and largely unpredictable market and regulatory forces, systematic evaluations will allow refinement and may avoid costly mistakes. In addition, developments in both earthquake and wind initiatives and the flood program suggest that an evaluation of the relationship between them and their interaction would be appropriate. Five years from enactment should provide enough time for the initiatives to become operational before they are evaluated" (Witt and Rubin, 1995: 17).

Consideration that regular evaluations form an integral part of any Canadian mitigation initiative will increase the chance of reaching objectives, and reduce the risk of unproductive activities becoming entrenched. Embracing evaluation is not characteristic of organizations, however, surmounting this tendency will be one of the overt characteristics of public sector managers, whose thinking is in the process of changing from a focus on survival to the attainment of excellence.

6. Conclusion: Strategic Considerations for Canada

Experience in Canada demonstrates a diversity of response capabilities and readiness within the institutional organizations responsible for emergencies. Continued focus on preparedness and response follows naturally from these established capabilities. but will show diminishing returns for future investments. The contradiction, here, is that beyond a certain level of preparedness and response capability, additional investment to further enhance response may save lives, but *only after lives have been lost and property damaged in the primary impact.* These underlying limitations of a response-oriented approach have drawn attention to the social character of disasters, and the potential of mitigation to realize loss reductions.

Choosing the road less travelled is not without risks, but then neither is standing still amid the currents of change. Undoubtedly the character of mitigation initiatives in Canada will differ from the evolving structure in the United States. To ignore the potential contribution of mitigation, and the added benefit of forging partnerships to share responsibilities and risk, would be imprudent in light of trends in losses, changes in the social composition of Canada, and the geographic distribution of the population.

This being said, the way forward must evolve in concert with a broad range of existing commitments and contentious jurisdictional issues. How then to proceed during a period of fiscal restraint, and with uncertainty of the benefits to be realized through mitigative actions? In response, natural disasters must be seen in their true societal context, and mitigation initiatives must be designed to reduce, not increase, overall expenditure. As with other emergency initiatives in Canada today, mitigation practices must be horizontally *and* vertically integrated* and incorporate iterative design features to achieve the loss reductions promised. A process based on thoughtful, reflective input combined with a willingness to share ideas and listen to others represents the best chance of achieving an effective, long-term mitigation strategy.

Further exploration of such a direction might productively employ a three-pronged concurrent approach, encompassing: (1) the development of a knowledge base for elected officials and citizens; (2) the generation of political will across jurisdiction boundaries; and (3) the encouragement of public-private partnerships. Entwining these three approaches will strengthen the concept, engage individuals and organizations, and begin construction of a sound foundation on which to build. With leadership and commitment, a mitigative attitude can evolve throughout government agencies and private organizations, that will result in the reduction of long-term risk to human life and property. The measure of success will not only be strong evidence of loss reductions, but also inclusion of mitigation activities in the everyday life of individuals, families, communities, and corporations.

In the years ahead, mitigation strategies must evolve to complement preparedness and response, reduce deaths and human suffering, realize reductions in loss expenditures by governments, insurers, businesses, and individuals, enhance the sustainability of settlements, and improve the quality of Canadians' lives. Moreover, the inevitability of natural disasters and the broad base of societal consequences, reinforces the value of developing public-private partnerships for the implementation of mitigation strategies to benefit all Canadians.

References

Bruce, J.: 1994, Introductory remarks, in J. McCulloch and D. Etkin (eds), *Proceedings of a Workshop on Improving Responses to Atmospheric Extremes: The Role of Insurance and Compensation*, Environment Canada, Downsview, pp. 1-1–1-5.

Burton, I.: 1996, Personal Communication, January 12th

Burton, I.: 1994, Costs of atmospheric hazards, in J. McCulloch and D. Etkin (eds), *Proceedings of a Workshop on Improving Responses to Atmospheric Extremes: The Role of Insurance and Compensation*, Environment Canada, Downsview, pp. 2-1–2-11.

Canadian National Committee for IDNDR: 1994, *National Report of Canada*, Royal Society of Canada/Canadian Academy of Engineering, Ottawa

deBeaupré, A.: 1996, Personal Communication, January 3rd.

Emergency Preparedness Canada (EPC): 1996a, *Federal Emergency Preparedness in Canada*, Draft, Unpublished.

Emergency Preparedness Canada (EPC): 1996b, *Civil Defense in Canada 1936–1972*, Draft, Unpublished.

* Horizontal integration refers to the coordinated effort of government departments and agencies at one jurisdictional level, as well as cooperation among private companies across different industrial/commercial sectors. Vertical integration refers to communication within national hierarchical structures, generally, but not exclusively intraministerial liaison. The overall objective is to enhance information flows and facilitate broadly based decision-making in a trans-organizational setting.

Emerson, W. and N. Mineta: 1995, *The Natural Disaster Protection Partnership Act – Legislative Outline*, Letter and attachments, June 6th.

Federal Emergency Management Agency (FEMA): 1995a, *National Mitigation Strategy – Partnerships for Building Safer Communities*, Mitigation Directorate, Washington, D.C.

Federal Emergency Management Agency (FEMA): 1995b, *Mitigation – Cornerstone for Building Safer Communities*, Mitigation Directorate, Washington, D.C.

Federal Emergency Management Agency (FEMA): 1994a, *Placing the Cornerstone – The Mitigation Directorate First-Year Report*, November 29, 1993–November 28, 1994, Mitigation Directorate, Washington, D.C.

Fowler, R. R.: 1993, *Report to Parliament on the Operation of the Emergency Preparedness Act, April 1, 1992–March 31, 1993*, Minister of Supply and Services, Ottawa.

Government of Canada: 1989a, *Canadian Civil Defense*, Fact Sheet, Emergency Preparedness Canada, Ottawa.

Government of Canada: 1989b, *Federal Emergency Policy*, Fact Sheet, Emergency Preparedness Canada, Ottawa.

Government of Canada: 1988a, *The Emergencies Act*, Chapter 29, pp. 777–818, Queen's Printer for Canada, Ottawa.

Government of Canada: 1988b, *The Emergency Preparedness Act*, Chapter 11, pp. 221–225, Queen's Printer for Canada, Ottawa.

Government of the United States: 1988, *Robert T. Stafford Disaster Relief and Emergency Assistance Act*, P.L. 93–288 as amended, Federal Emergency Management Agency, Washington, D.C.

Hewitt, K.: 1983, *Interpretations of Calamity – from the Viewpoint of Human Ecology*, Allen & Unwin Inc., Boston.

Hewson, M. D.: 1996, Personal Communication, January 4th.

Interagency Floodplain Management Review Committee (IFMRC): 1994, *Sharing the Challenge: Floodplain Management into the 21st Century*, (The Galloway Report), United States Government Printing Office, Washington, D.C.

International Federation of Red Cross and Red Crescent Societies (IFRC/RCS): 1993, *World Disaster Report*, Martinus Nijhoff Pub., The Netherlands.

Jacobs, J.: 1992, *Systems of Survival – A Dialogue on the Moral Foundations of Commerce and Politics*, Random House, New York.

Lechat, M. F.: 1990, 'The International Decade for Natural Disaster Reduction: Background and Objectives', *Disasters*, **14**(1), 1–6.

Michleburg, R.: 1996, China wary of economic reform, *The Globe and Mail*, February 19: A10.

Mileti, D. S. *et al.*: 1995, Towards an integration of natural hazards and sustainability, *The Environmental Professional*, **17**(2), 117–126.

Moore, R. T.: 1996, Personal Communication, January 19th.

Myers, M. F.: 1996, Personal Communication, January 26th.

National Research Council: 1994, *Facing the Challenge – The U.S. National Report to the IDNDR World Conference on Natural Disaster Reduction*, National Academy Press, Washington, D.C.

Neville, D. C.: 1996, Personal Communication, January 3rd.

Newton, J.: 1995, *Coping in Context – Adaptation to Environmental Hazards in the Northern Regions of Canada*, Ph.D. Dissertation, Department of Geography, University of Toronto.

Peters, D. W.: 1996, Personal Communication, January 3rd.

Platt, R.: 1996, 'Hazard Mitigation: Cornerstone or Grains of Sand?', *Disaster Research 202*, Boulder, Colorado, Item 1.

Quarantelli, E. L.: 1989. Planning and management for the prevention and mitigation of natural disasters, especially in a metropolitan context: Initial questions and issues which need to be addressed, *Planning for Crisis Relief, Vol. 3 Planning and management for the Prevention and Mitigation from Natural Disasters in Metropolis*. pp. 1–17, Nagoya, Japan.

Ross, A.: 1994, The Canadian insurance industry, in J. McCulloch and D. Etkin (eds.), *Proceedings for a Workshop on Improving Responses to Atmospheric Extremes: The Role of Insurance and Compensation,* Environment Canada, Toronto, pp. 4-1–4-7.

Ross, A.: 1996, *Climatic Change and its Impact on the Canadian Insurance Industry*, Presentations at the Institute of Environmental Studies, University of Toronto, February 19, 1996.

Rossi *et al.*: 1983, *Victims of the Environment – Loss from Natural Hazards in the United States, 1970–1980*, Plenum Press, New York.

Rubin, C. B. (ed).: 1985, *Community Recovery from a Major Natural Disaster*, Institute of Behavioural Science, Boulder, Colorado.

Shea, R. F. Jr.: 1996, Personal Communication, January 19th.

Showalter, P. Sands, Riebsame, W. E., and Myers, M. F.: 1993, *Natural Hazards Trends in the United States: A Preliminary Review for the 1990s*, Institute of Behavioural Science, University of Colorado, Boulder.

Simard, A. J.: 1996, Personal Communication, January 4th.

Treacy, M. and Wiersema, F.: 1995, *The Discipline of Market Leaders*, Addison-Wesley Pub. Co., Reading, Massachusetts.

Tsai, F. Y.: 1996, Personal Communication, January 19th.

Tubbesing, S. K.: 1989, 'Natural Hazard Reduction in the United States: A Brief Assessment', *Report of the Colorado Workshop on Hazard Mitigation in the 1990's*, Natural Hazards Research and Application Information Center, Boulder, pp. 35–37.

United Nations Department of Humanitarian Affairs: 1992, *Internationally Agreed Glossary of Basic Terms related to Disaster Management*, DNA/93/36, Geneva.

Vollard, R. H.: 1996, Personal Communication, January 18th.

Whiteley, R. C.: 1991, *The Customer Driven Company – Moving from Talk to Action*, The Forum Corporation, Addison-Wesley Pub. Co., Reading, Massachusetts.

Witt, J. L.: 1995, *Presentation at National Mitigation Conference*, December 6, 1995, Arlington, Virginia.

Witt, J. L. and Rubin, R. E.: 1995, *Administration Policy Paper – Natural Disaster Insurance and Related Issues*, Federal Emergency Management Agency and Department of the Treasury, Washington, D.C.

Natural Hazards **16:** 243–265, 1997.

Arias Intensity Assessment of Liquefaction Test Sites on the East Side of San Francisco Bay Affected by the Loma Prieta, California, Earthquake of 17 October 1989

ROBERT E. KAYEN[1] and JAMES K. MITCHELL[2]
[1]*Research Civil Engineer, U.S. Geological Survey, 345 Middlefield Road, Menlo Park, CA 94025 U.S.A. rkayen@usgs.gov*
[2]*University Distinguished Professor of Civil Engineering, Virginia Tech, 109B Patton Hall, Blacksburg, VA 24061-0105 U.S.A. e-mail: jkm@vk.edu*

(Received: 25 February 1997; in final form: 2 September 1997)

Abstract. Uncompacted artificial-fill deposits on the east side of San Francisco Bay suffered severe levels of soil liquefaction during the Loma Prieta earthquake of 17 October 1989. Damaged areas included maritime-port facilities, office buildings, and shoreline transportation arteries, ranging from 65 to 85 km from the north end of the Loma Prieta rupture zone. Typical of all these sites, which represent occurrences of liquefaction-induced damage farthest from the rupture zone, are low cone penetration test and Standard Penetration Test resistances in zones of cohesionless silty and sandy hydraulic fill, and underlying soft cohesive Holocene and Pleistocene sediment that strongly amplified ground motions. Postearthquake investigations at five study sites using standard penetration tests and cone penetration tests provide a basis for evaluation of the Arias intensity-based methodology for assessment of liquefaction susceptibility.

Key words: acceleration, arias intensity, earthquake, geotechnical, liquefaction, field test, penetration resistance, seismometer.

1. Introduction

The 1989 Loma Prieta earthquake ($M_s = 7.1$) occurred when a segment of the San Andreas fault northeast of Santa Cruz, Calif. ruptured over a length of approximately 45 km (Borcherdt, 1994; Spudich, 1996). The hypocenter which was located at lat. 37.037 N. and long. 121.883° W. at 18 km depth, and during the next 8 to 10 s ruptured bilaterally for approximately 20 km northward and 20 km southward. The rupture propagated toward the Earth's surface, up to a depth of approximately 5 to 7 km.

Peak horizontal accelerations on sites underlain by rock and stiff alluvium in the east bay generally ranged from 0.08 to 0.12 g, but amplification due to the presence of soft and deep cohesive soil deposits underlying artificial fills produced peak accelerations from 0.11 to 0.29 g at strong-motion recording stations located on these fills (Figure 1). Peak accelerations near most of our study sites can be estimated from strong motion records obtained near the east bay shoreline. On the basis of available strong motion data (Shakal *et al.*, 1989; Kayen *et al.*, 1992), the peak horizontal accelerations on bayshore fills in the vicinity of Oakland International Airport, and Bay Farm Island were apparently about 0.27 g, and at the San Francisco–Oakland Bay Bridge toll plaza and Port of Oakland's container terminal at Seventh Street were probably about 0.28 to 0.29 g (Figure 1). Arias intensities, integrated from the strong motion records ranged from 0.8 m/s at Oakland International Airport, Bay Farm Island, and Alameda Naval Air Station to 1.71 m/s at the Port of Oakland's container terminal at Seventh Street and the Bay Bridge toll plaza (Kayen and Michell, 1997).

At the Port of Richmond, the most northerly study site, no nearby strong-motion recordings were obtained for sites with similar soil conditions (Figure 1). Approximately 2 km to the north, however, a peak horizontal acceleration of 0.13 g on the south component was recorded at Richmond's City Hall. We used this record, the CSMIP strong motion recording Station No. 58505, to model the site response at the Port of Richmond study site where soft, deep cohesive-soil deposits underly artificial fill, by means of one-dimensional equivalent-linear seismic site-response analyses, using the computer program SHAKE90 (a modified version of the computer program SHAKE; Schnabel *et al.*, 1972). These results indicate that peak horizontal accelerations at the Port of Richmond were probably about 0.18 g. Synthetic seismograms from the SHAKE90 analysis were integrated and we estimate that the Arias intensity at the Port of Richmond site to be approximately 0.6–0.7 m/s.

This paper describes soil liquefaction, ground deformations and associated liquefaction-damages in artificial fill deposits on the east-bay shoreline from Oakland International Airport to the Port of Richmond, and presents the results of our post earthquake studies at five study sites near the east bay shoreline: (1) Oakland International Airport (Figures 1 and 2), (2) Bay Farm Island and Harbor Bay Plaza (Figures 1 and 2), (3) the Port of Oakland's Marine Container terminal at Seventh Street (Figures 1 and 3), (4) the San Francisco–Oakland Bay Bridge Toll Plaza (Figures 1 and 3), and (5) the Port of Richmond (Figures 1 and 4). The collapse of the Interstate Highway 880 Cypress Street Viaduct occurred near sites 3 and 4, along a stretch of viaduct that is founded on Holocene Bay Mud. Liquefaction was not observed at this site and so we do not discuss damage to the Cypress Freeway here. Detailed discussion of the geotechnical aspects of the Cypress Freeway collapse can be found in Seed *et al.* (1990).

2. Methods

Standard penetration tests (SPT) were performed following the guidelines of the American Society for Testing and Materials (1984). The electronic cone-penetration-tests (CPT) followed the procedures of the American Society for Testing and Materials (1986). To remove the influence of effective overburden stress on

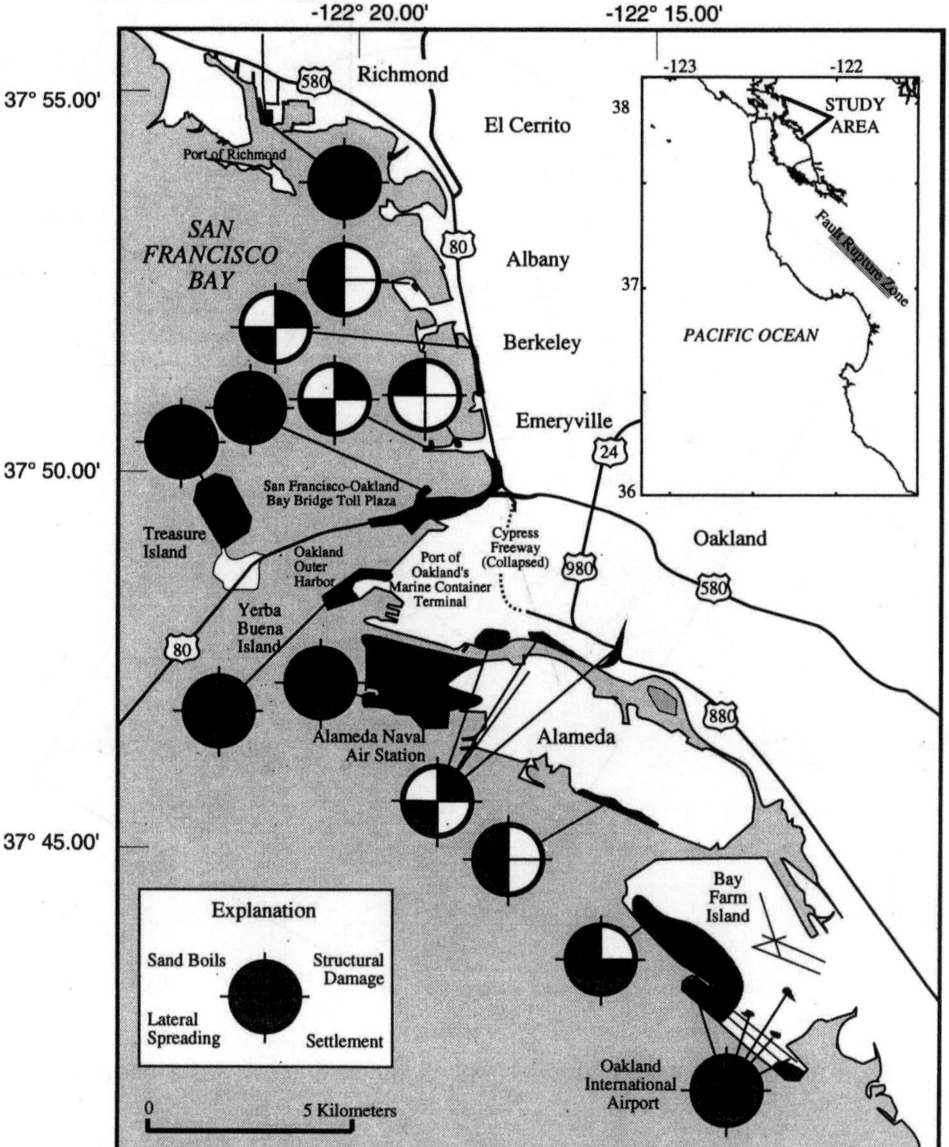

Figure 1. San Francisco east-bay shoreline, showing locations of sites where liquefaction-induced damage occurred from the 1989 Loma Prieta earthquake and locations of study sites where soil investigations were performed.

the field penetration resistance measurements so that we may evaluate the liquefaction potential of soil in a manner that is independent of soil depth, we normalized the field measurements to values at a corresponding reference stress. For example, N values, taken during SPT tests, can be normalized to a reference effective over-

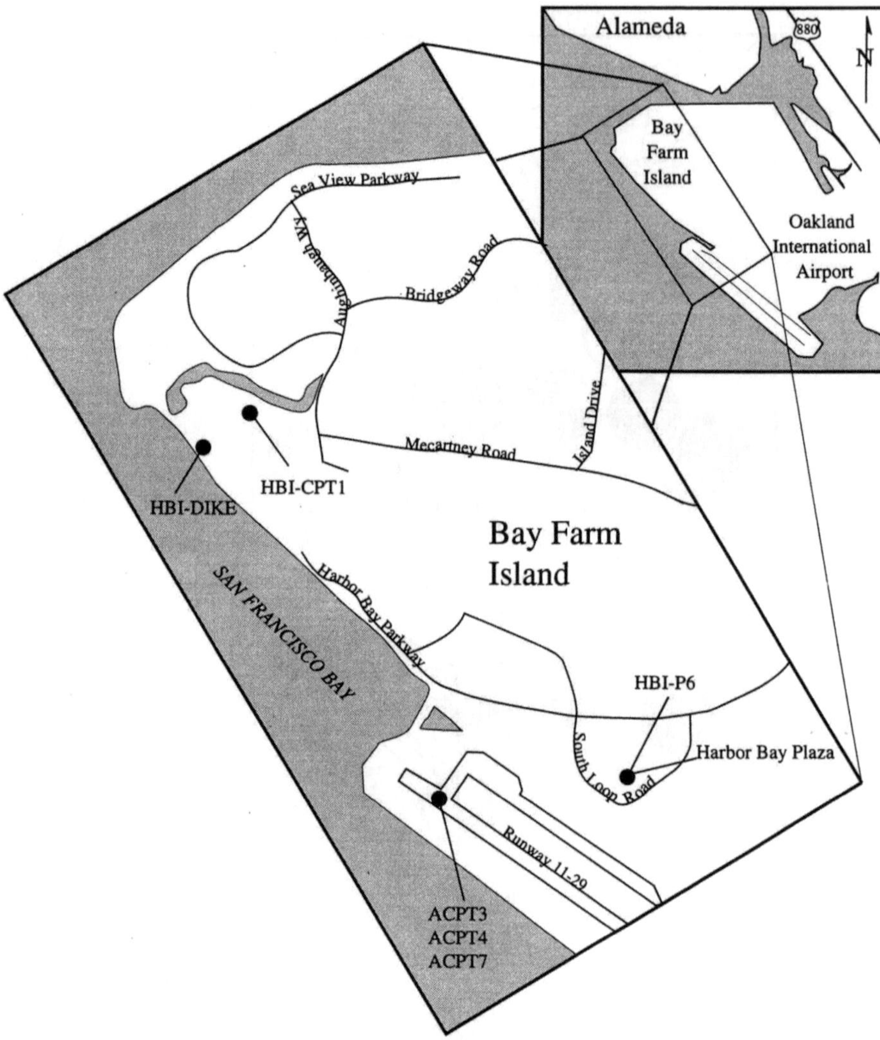

Figure 2. The northern part of Oakland International Airport and Bay Farm Island, showing locations of boreholes (dots).

burden stress of 1 atm (σ'_{vo} = 0.096 MPa) by correcting for overburden effects on penetration resistance, using the relation of Seed and Idriss (1971)

$$N_1 = C_n N. \tag{1}$$

We found that the overburden-stress-correction factor, C_n, can be reasonably expressed as

$$C_n = \frac{2.2}{1.2 + \left(\frac{\sigma'_{vo}}{\sigma_{\text{ref}}}\right)} \tag{2}$$

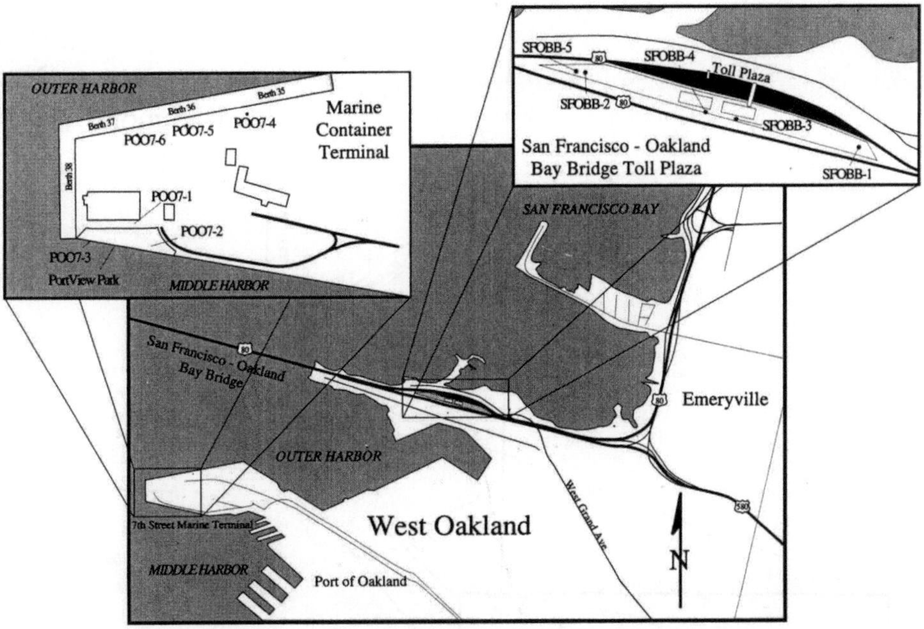

Figure 3. Emeryville–Oakland shoreline, showing locations of boreholes (dots) near San Francisco–Oakland Bay Bridge Toll Plaza and Port of Oakland Marine Container Terminal at Seventh Street.

where N_1 is the overburden-stress-corrected standard penetration resistance, σ'_{vo} is the effective overburden-stress, and σ_{ref} is a reference stress of 1 atm. Likewise, to evaluate q_c values measured during CPT tests we adjust the measured q_c values as follows:

$$q_{c1} = C_q q_c. \tag{3}$$

We found that C_q can be reasonably expressed as

$$C_q = \frac{1.8}{0.8 + \left(\dfrac{\sigma'_{vo}}{\sigma_{\text{ref}}}\right)}. \tag{4}$$

Liquefiable layers at the Toll Plaza, Port of Oakland, Bay Farm Island, and Oakland International Airport study sites can be characterized as clean sand with a mean grain diameter (D_{50}) of at least 0.25 mm and a fine-particle content of less than 5%. The liquefiable layer at the Port of Richmond study site has a D_{50} value of approximately 0.06 to 0.07 mm and a fine-particle content of >35%. Silty sands, like those at the Port of Richmond site, with the same liquefaction resistance to that of a corresponding reference 'clean' sand are typically observed to have distinctly lower measured penetration resistance values. Seed *et al.* (1984)

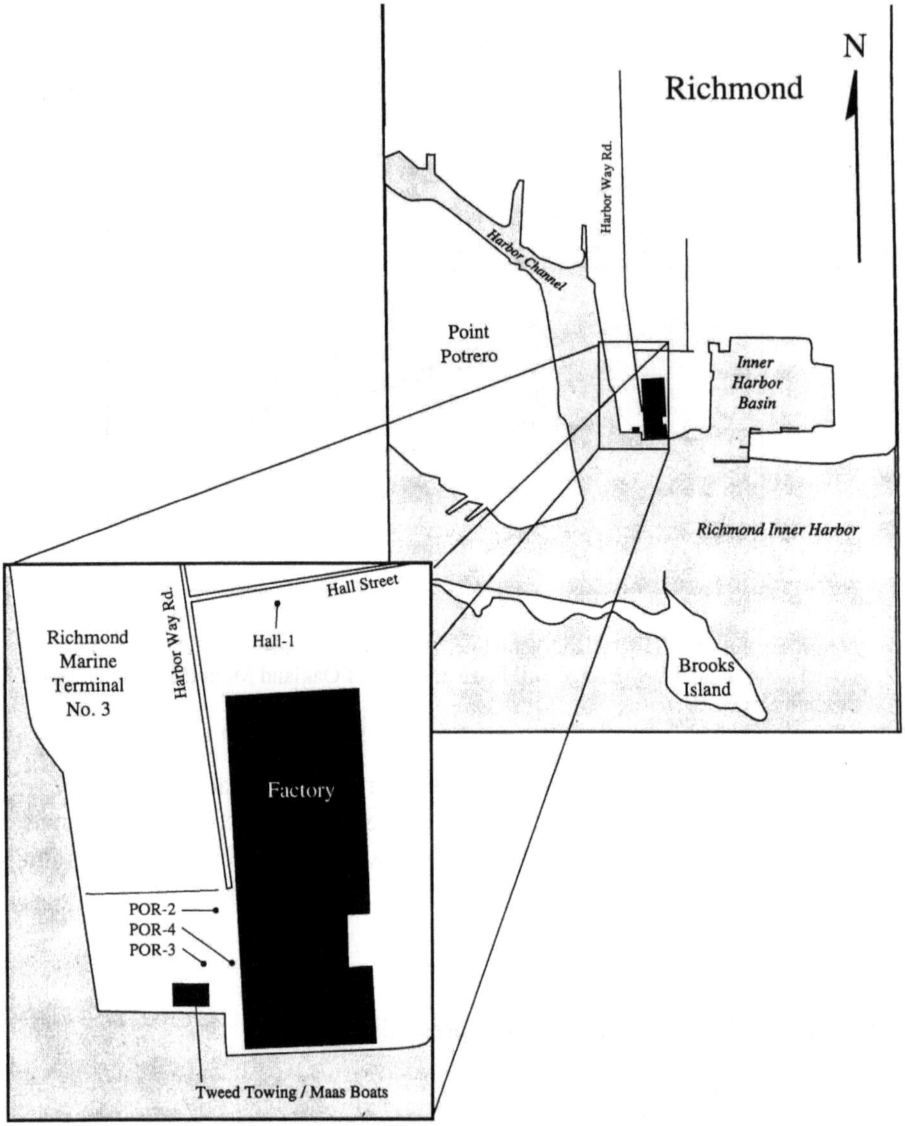

Figure 4. Port of Richmond and Inner Harbor Basin, showing locations of boreholes (dots).

presented boundary curves that effectively convert penetration resistance values of silty sands to that for clean sands with no fine particles, and defined a SPT-fines content correction factor (Seed and de Alba, 1986):

$$(N_1, f_c)_{\text{clean sand equivalent}} = (N_1)_{\text{silty sand measured}} + \Delta N_1. \quad (5)$$

The fines-correction factor ΔN_1 is a function of the fines content, FC, in percent. Following the guidelines presented in the consensus statement from the NCEER

Workshop on Evaluation of Liquefaction Resistance, 4–5 January 1996, Salt Lake City, Utah, we used the following equations for correcting SPT-values of low-plasticity silty sands to equivalent values for clean sand

$$\Delta N_1 = \begin{pmatrix} \text{FC} \leq 5\%: & 0 \\ 5\% < \text{FC} < 35\%: & (FC - 5)(7/30) \\ \text{FC} \geq 35\%: & 7 \end{pmatrix} \quad (6)$$

We applied these correction factors to the Port of Richmond site SPT data. Detailed description of the borehole logs from all of the liquefaction test sites, including those discussed herein, can be found in Kayen *et al.* (1992, 1997), Kayen (1993) and Mitchell *et al.* (1994).

3. Field Investigations

Oakland International Airport: Soil liquefaction caused considerable damage to the main jet runway (11–29) at Oakland International Airport (Figure 2). Additional evidence of liquefaction, including sand boils, settlement and lateral spreading, occurred over wide areas of airport fill to the north, south and east of the damaged runway section. Much of the runway and inboard taxiway area is built on loose, sandy fill underlain at shallow depths by soft bay mud. The perimeters of the airport fill have dikes to prevent inundation during unusually high tides and storms.

Extensive soil liquefaction occurred in the western section of the airport fill, damaging the northwestern 900 m of the 3000 m-long main runway. The adjacent taxiway pavement was also heavily damaged. Most of the runway damage was repaired within 4 weeks, and as a result, the airport was able to resume essentially full operations with a shortened operational runway 2700 m long on 20 November 20 1989.

In addition to liquefaction, settlement, and lateral spreading of the main runway fill, settlement and lateral spreading also occurred in several places in the surrounding perimeter dikes at the west end of the runway fill. The maximum observed levee settlement of the perimeter dikes was approximately 0.5 to 0.7 m, and lateral deformations were similar in magnitude. Liquefaction was also observed near the main terminal buildings.

The logs of three boreholes drilled at the Airport show little or no surficial crust between 0 and 2 m depth, where the water table was penetrated (log ACPT7 is shown in Figure 5). From 2 to 4.5–5.8 m depth, is an extremely loose deposit of fine-sand hydraulic fill that almost certainly was responsible for the observed liquefaction and lateral spreading. q_c values within this deposit range from 2 to 14 MPa, and N values from 1 and 4 blows/ft. The material in this layer correlates with sand boil material observed at the surface.

Bay Farm Island and Harbor Bay Plaza: At Bay Farm Island, immediately north of Oakland International Airport, considerable liquefaction occurred at the northwest corner and at points along the west edge of the fill (Figures 1 and 2). Most

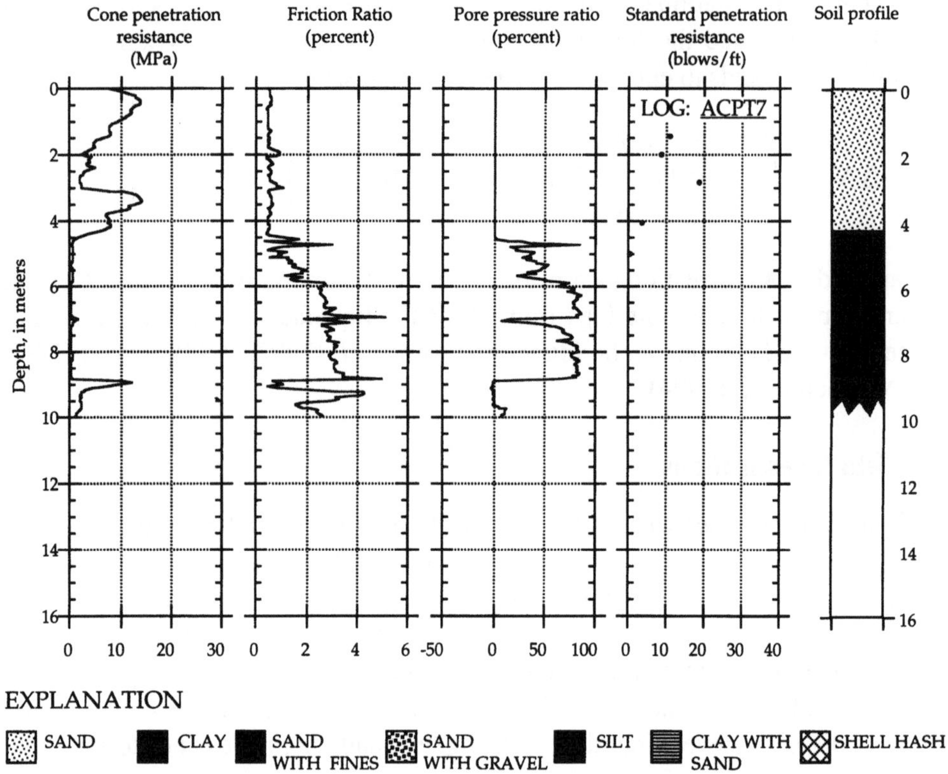

Figure 5. Log of borehole ACPT7 in fill deposits at Oakland International Airport test site (see Figure 2 for location).

of the western part of Bay Farm Island consists of sandy hydraulic fill, underlain by bay mud and deeper, stiffer alluvium and estuarine sediment. Soil liquefaction on the island was largely confined to undeveloped, artificially filled land. Numerous sand boils, many relatively large (with diameters more than 3 m) occurred in this area. Fill in a perimeter dike that bounds the western side of Bay Farm Island was densified by dynamic compaction, which appears to have successfully prevented soil liquefaction of the dike during the earthquake.

The logs of three boreholes drilled at the Bay Farm Island study sites suggest a complex stratigraphy of relatively thin (less than 1 m thick) interbeds of sand and finer material. A total of 20 CPT logs were also taken at one of the liquefaction test sites to observe any changes in strength over time; a representative log (HBI-CPT1) is presented in Figure 6. The logs in the unimproved fill indicate a noncohesive layer of sand and silty sand at 2 to 3 m, with q_c values of typically 2 to 16 MPa and N values of 8 to 18 blows/ft. The water table is within this layer at between 1.5 and 2 m depth. Below this layer, to approximately 6.5 m depth, is a set of finer interlayered deposits exhibiting decreasing q_c values, and elevated B_q and FR values. Below 6.5 m depth, silty sand q_c defines the rest of the logged soil column.

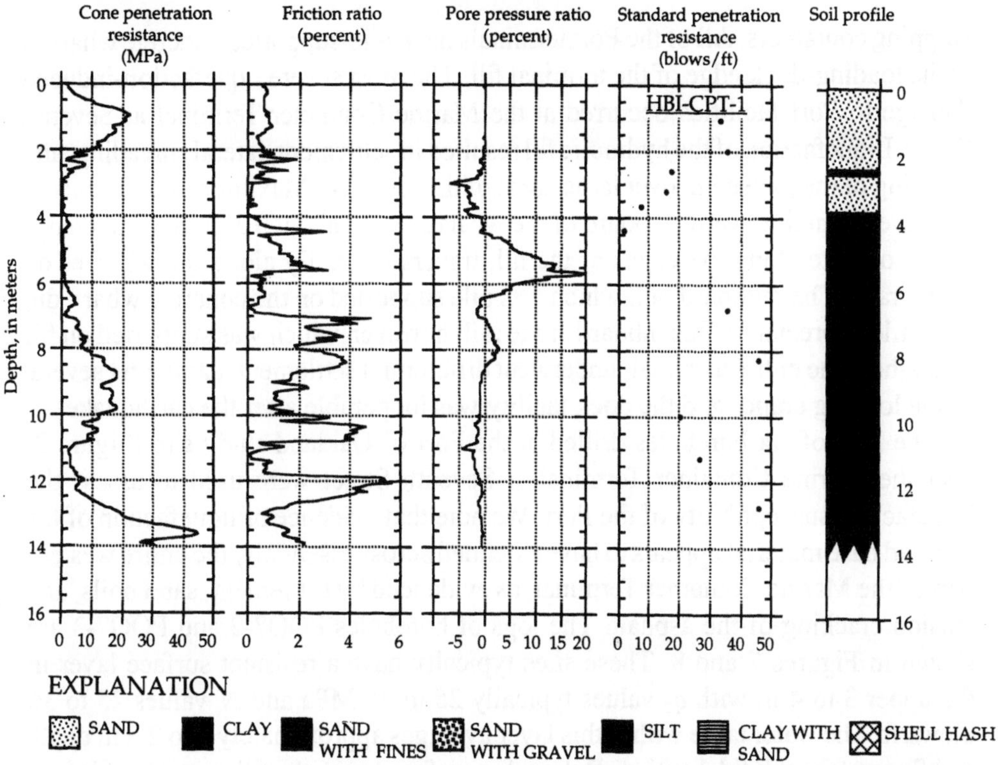

Figure 6. Log of borehole HBI-CPT1 in fill deposits at Bay Farm Island test site (see Figure 2 for location).

Liquefaction probably occurred in the fine sand and silty sand at 2 to 3 m depth. on the basis of correlation of sand boil materials with recovered samples.

CPT and SPT logs were taken along the improved perimeter dike, which did not liquefy during the earthquake. At the elevated dike structure, the water table is somewhat deeper in the soil profile at 2.5 m depth. However, within the sandy deposits at 2.5 to 7 m depth, q_c values are noticeably higher, ranging from 10 to 35 MPa, and N values of 29 to 58 blows/ft.

The Port of Oakland: Immediately south of the San Francisco–Oakland Bay Bridge, soil liquefaction caused considerable damage to marine-container facilities at several places in the Port of Oakland, adjacent to the Oakland Outer Harbor, as shown in Figures 1 and 3. Much of the extreme western part of the area near the Oakland shoreline, south of the mole, is filled land underlain by a relatively thin layer of soft, normally consolidated Holocene marine clay (bay mud). These shallow surficial units are, in turn, underlain by upwards of 180 m of stiffer Pleistocene glacial and interglacial deposits.

Most of the surface fill at the Port of Oakland was hydraulically placed to sea level, above which a combination of hydraulic and dumped fill was placed. After placement of the unimproved hydraulic fill, a relatively thick asphalt-cement pavement was laid in many of the terminal areas to support heavy vehicles and shipping containers. All of the Port terminals have pile-supported concrete wharves at the loading-dock edge of the terminal fill. The most severe liquefaction-induced damage to Port facilities occurred at the Marine Container Terminal at Seventh Street. Liquefaction of the hydraulic fill resulted in settlement, lateral spreading, and cracking of the pavement over large areas of the terminal. Maximum settlements of the paved container yards inboard of the wharves were about 0.3 m. Several cranes which operate along the edges of the fill, traverse laterally along the wharves on heavy rails. The outboard rail, which was pile supported on the concrete wharf, did not settle appreciably. The inboard crane rail, however, which was supported on fill throughout the entire terminal, underwent differential settlement, damaging several of the loading cranes and the dock facility was inoperable after the earthquake.

The logs of six boreholes drilled at the Port of Oakland study site (Figure 2) near the Marine Container Terminal at Seventh Street were used to assess the liquefaction susceptibility of the Port. We note that evidence of liquefaction of the subroad base material appears to have extended across essentially the entire western part of the Marine Container Terminal, as evidenced by settlement, sand boils, and tension cracking of the asphalt. The logs of boreholes POO7-2 and POO7-3 are shown in Figures 7 and 8. These sites typically have a resistant surface layer in the upper 3 to 4 m with q_c values typically 25 to 35 MPa and N values 25 to 36 blows/ft. The water table within this layer averages approximately 2 to 2.3 m depth and fluctuates with tidal action. Below this surface layer, the fill consists of looser deposits of fine marine sand with q_c values of 8 to 15 MPa, and N values of 10 and 25 blows/ft. Liquefaction appears to have occurred in the materials with in the depth range 4–8 m, on the basis of correlations of SPT samples with surface sand-boil material recovered.

San Francisco–Oakland Bay Bridge Toll Plaza: The San Francisco–Oakland Bay Bridge mole (peninsula-approach fill), immediately south of Emeryville, was extensively damaged by soil liquefaction (Figures 1 and 3). Appreciable settlement (max. 40 cm) occurred over most of the mole. Lateral spreading also was significant along most of the fill, causing numerous longitudinal fissures in the pavement parallel to the fill edges. Settlements of the mole and the elevated structures of Interstate Highway 580 and West Grand Avenue were severe, resulting, at each site, in pavement collapse and open fissures. as much as 0.3 m wide, at the soil-structure interface. The approach mole also settled below the bridge- and ramp-structure road level by as much a 0.5 m. The fissures and uneven pavement surface interrupted ground transportation in the interchange areas after the earthquake.

The logs of five boreholes drilled at the San Francisco–Oakland Bay Bridge Toll Plaza study site indicate a complex stratigraphy. Five sets of logs were taken at locations along the median strip and open space between the east bound and west bound lanes of Interstate Highway 80. The log of borehole SFOBB-1 was taken at the easternmost end of the Toll Plaza parking lot. The log is marked by a moderately dense surface crust between 0 and 2.8 m depth, followed by an apparently interbedded sequence of cohesionless silty sand and cohesive clay to

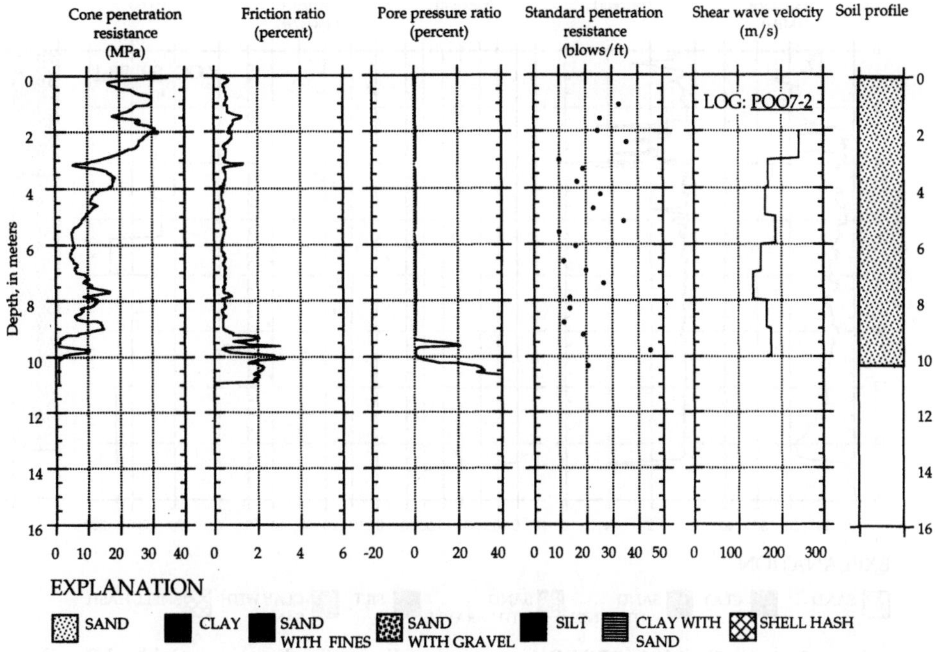

Figure 7. Log of borehole POO7-2 in fill deposits at Port of Oakland (see Figure 3 for location).

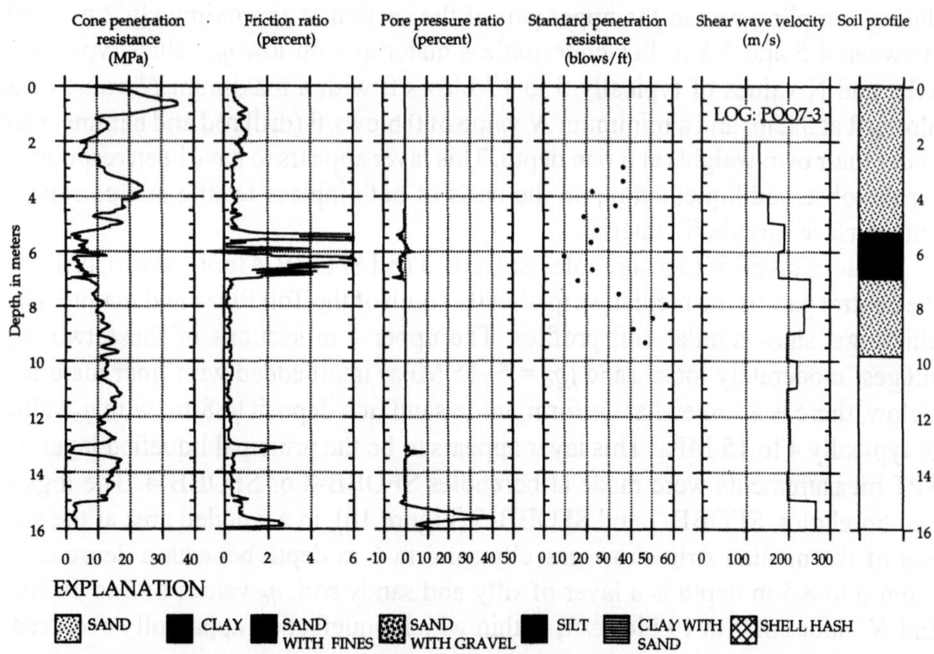

Figure 8. Log of borehole POO7-3 in fill deposits at Port of Oakland (see Figure 3 for location).

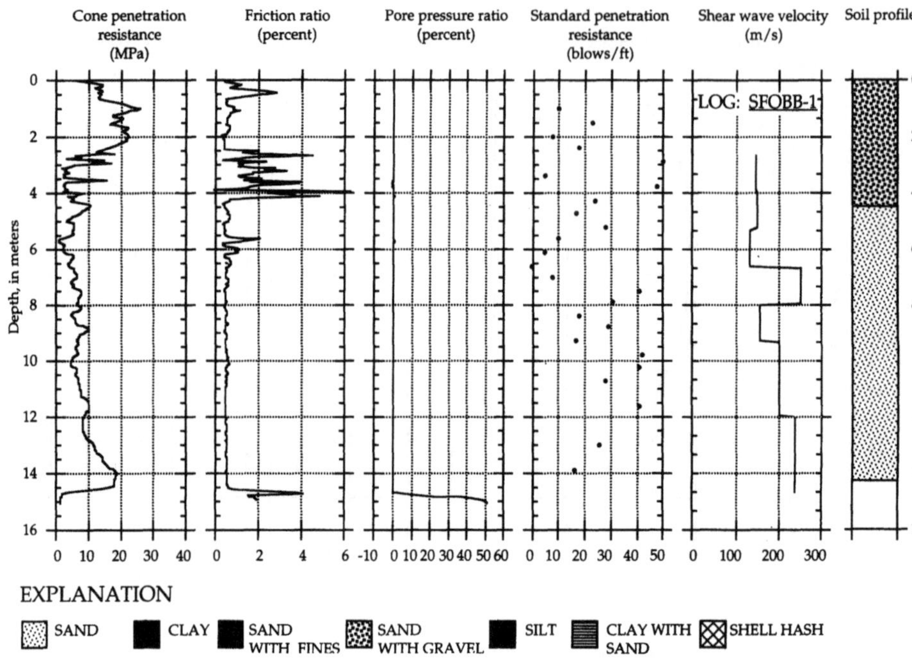

Figure 9. Log of borehole SFOBB-1 on east shoreline of San Francisco–Oakland Bay Bridge mole (see Figure 3 for location).

4.5 m depth, as noted by the local variations in FR and q_c values. The water table during sampling was in the upper part of the section at approximately 2 m depth. Between 4.5 and 7.5 m lies cohesionless material with low q_c values typically 5 MPa and N values of typically 5 to 17 blows/ft with a maximum N value of 28 blows/ft at 5.2 m, and a minimum N value of 0 blows/ft (drill rod and hammer sank under their own weight) at 6.6 m depth. This layer appears to have been responsible for the observed liquefaction, on the basis of both inplace testing and correlation with surface sand-boil material.

The logs taken of two boreholes SFOBB-3 and SFOBB-4 (not shown), drilled in the central part of the median strip directly south of the Toll Plaza and maintenance buildings, show similar soil profiles. The upper 4 m sections of these two logs suggest moderately loose sand (q_c = 5–15 MPa) interbedded with finer materials. Below, there is a somewhat uniform, more sand rich deposit to 8-m, with q_c values of typically 4 to 15 MPa. This layer appears to be the principal liquefied layer. No SPT measurements were made at boreholes SFOBB-3 or SFOBB-4. The logs of two boreholes, SFOBB-2 and SFOBB-5 (Figure 10), in a wooded area at the west end of the median strip show gravelly sand to 6 m depth beneath a dense crust. From 6 to 8.5 m depth is a layer of silty and sandy soil, q_c values of 5 to 10 MPa and N values of 7 to 19 blows/ft, within which liquefaction apparently occurred.

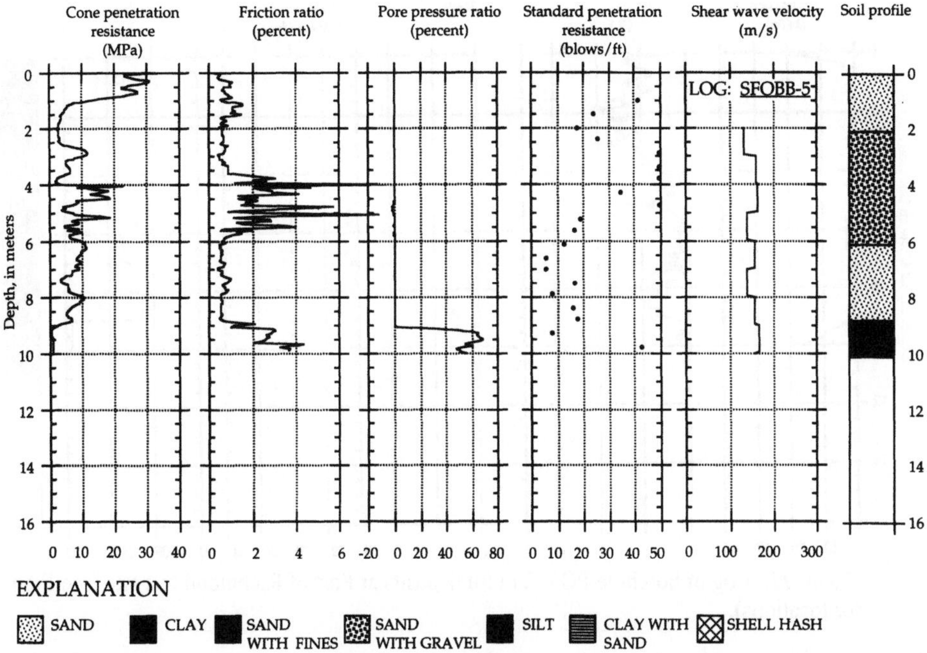

Figure 10. Log of borehole SFOBB-5 on east shoreline of San Francisco–Oakland Bay Bridge mole (see Figures 3 for location).

Port of Richmond: Soil liquefaction occurred at a study site in the western part of Richmond Inner Harbor (Figures 1 and 4). The area that liquefied is a zone approximately 75 m wide and 300 m long at the foot of Harbor Way Road. (10th Street). Approximately 85 km north of the Loma Prieta rupture zone, this site represents the most distant point from the zone of energy release to undergo soil liquefaction sufficient to damage structures. Much of the land at the edge of the Richmond Inner Harbor had been created by placement of uncompacted sandy hydraulic fill. The liquefaction zone overlies deposits of soft bay mud, which are underlain, in turn, by deeper deposits of stiffer, overconsolidated noncohesive and cohesive soils that naturally fill a deep fluvial channel at the west end of the harbor. Four large and a dozen smaller sand boils vented fine sand and silty sand from the underlying fill. In addition, minor settlements of approximately 2 to 8 cm and fissures of similar magnitude occurred at the edge of the harbor.

Three boreholes (POR-2 through POR-4) drilled in the open space north of the Harbor seawall (Figures 1 and 4). SPT's were performed in shallow boreholes adjacent to each CPT. The logs of the boreholes drilled at the Port of Richmond study site (Figure 2) are all similar and are characterized in the upper part of the section by an oxidized, tan-brown crusted silty sand to 0.8 m depth; oxidized tan-brown silty sand to 1.8 m depth; and oxidized, tan-brown sandy clay to reduced sandy clay to 4.1 m depth. The water table was approximately at a depth of 2.5 m during sampling. This section has low q_c values of approximately 4 to 5 MPa, FR

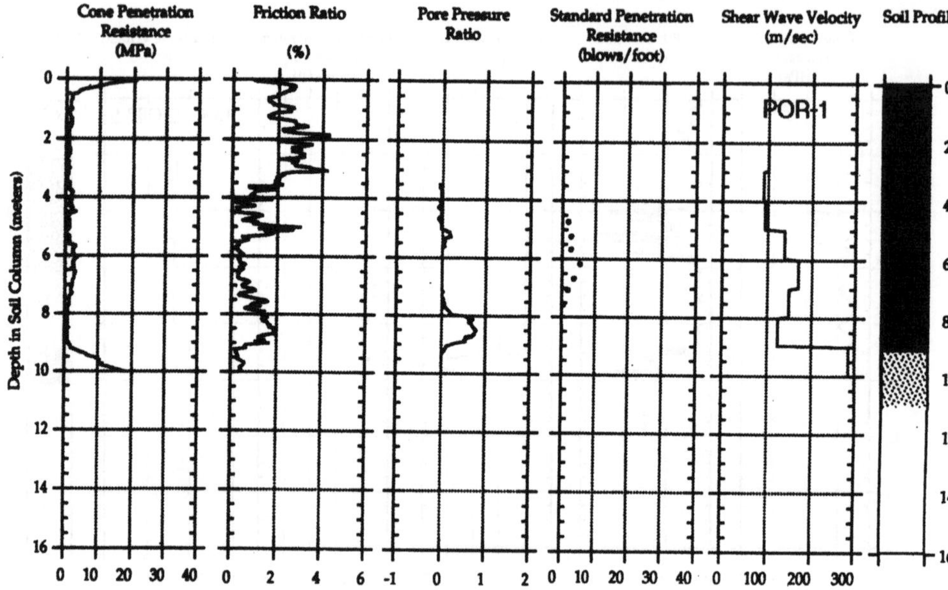

Figure 11. Log of borehole POR-2 in fill deposits at Port of Richmond test site (see Figure 4 for locations).

values greater than 3% and pore pressure generation, and N values ranging from 1 to 6 blows/ft. Below this section, at 4.1 to 7.8 m depth, is a hydraulic-fill layer of olive gray fine-silty sand containing shell fragments. This layer almost certainly was responsible for the observed liquefaction at the surface, and surface sand-boil material is identical to SPT samples of this layer. This layer has extremely low q_c values 1 to 3 MPa, average FR values of 0.3 to 1.0%, and minor pore pressure generation during cone penetration, and N values of 2 to 11 blows/ft, but typically, 2 to 5 blows/ft. Below this layer is a thin deposit of soft bay mud to 9.5 m depth, underlain by dense sand deposits.

4. Discussion

We compared both the observed liquefaction behavior and the liquefaction potential of the study sites, as measured by q_c and N, with the measured earthquake shaking intensity recorded at nearby seismometer sites. Two approaches were used to assess the intensity of seismic shaking: the first approach used the conventional cyclic stress method (CSR) of Seed and others (1984); and the second approach utilizes a new method based on the Arias intensity of ground motion recorded near the liquefaction study sites (Kayen and Mitchell, 1997). The first method utilizes a simplified uniform cyclic stress ratio to represent the complex and irregular earthquake-induced stress-time history. This equivalent series of cyclic loads of uniform amplitude is expressed as follows (Seed and Idriss, 1982)

$$\mathrm{CSR} = 0.65 \frac{a_{\max}}{g} \frac{\sigma_{vo}}{\sigma'_{vo}} r_d, \qquad (7)$$

where a_{max} is the peak ground acceleration, g is the gravitational acceleration (9.81 m/s^2), σ_{vo} is the total overburden stress, σ'_{vo} is the effective overburden stress, and r_d is a depth-reduction factor, which can be estimated in the upper 10 m of the soil column by the following relation:

$$r_d = 1 - 0.012z. \tag{8}$$

where z is the depth in meters (Kayen *et al.*, 1992).

The second approach utilizes a quantitative measure of earthquake shaking intensity, termed Arias intensity (Arias, 1970). It is the sum of the energy absorbed by an evenly-spaced population of idealized undamped simple oscillators in response to the earthquake motion. For the two horizontal components of motion, the Arias intensity I_h is calculated as follows:

$$I_h = I_{xx} + I_{yy} = \frac{\pi}{2g}\int_0^{t_0} a_x^2(t)\,\mathrm{d}t + \frac{\pi}{2g}\int_0^{t_0} a_y^2(t)\,\mathrm{d}t, \tag{9}$$

where I_{xx} is the Arias intensity measured in the x direction; I_{yy} is the Arias intensity in the y direction; g is the acceleration due to gravity; t_0 is the duration of earthquake shaking; and a_x and a_y are the transient accelerations in the x and y directions. The Arias intensity measure has the dimensional units of velocity.

As in the case of a_{max}, we find that I_h typically diminishes with depth in the soil column. For liquefaction assessment we need to know the I_h profile within the soil column. We estimated the attenuation of I_h with depth through a parametric study of synthetic accelerograms propagated through shear modulus representations of soil columns at our east bay study sites. Synthetic seismograms were generated for the surface and depth nodes of soil column input files with the ground-response computer-program SHAKE90, a modified version of the program presented by Schnabel *et al.* (1972). Strong motion records from eight earthquakes were propagated through a soil column representing loose sandy fill (V_s = 150 m/s.) overlying a cohesive soil with shear wave velocity profiles for San Francisco bayshore sites (Sun *et al.*, 1989; Golesorkhi, 1989; Kayen, 1993). Each profile was underlain by an elastic half-space with a shear wave velocity of 2500 m/s, representative of the Mesozoic Franciscan Formation underlying the east bay sites. Output synthetic acceleration-time histories for layer nodes at depth in the model soil profiles were integrated to calculate I_h and then normalized by the value of I_h at the surface. This normalization process allows us to evaluate the depth-dependency of I_h over a broad range of earthquake and site conditions by collapsing the profiles to a common reference value (unity) at the ground surface. We define the Arias intensity depth-of-burial reduction parameter, r_b, as the ratio of the buried-to-surface Arias intensity.

$$r_b = \frac{I_h(\text{depth})}{I_h(\text{surface})}. \tag{10}$$

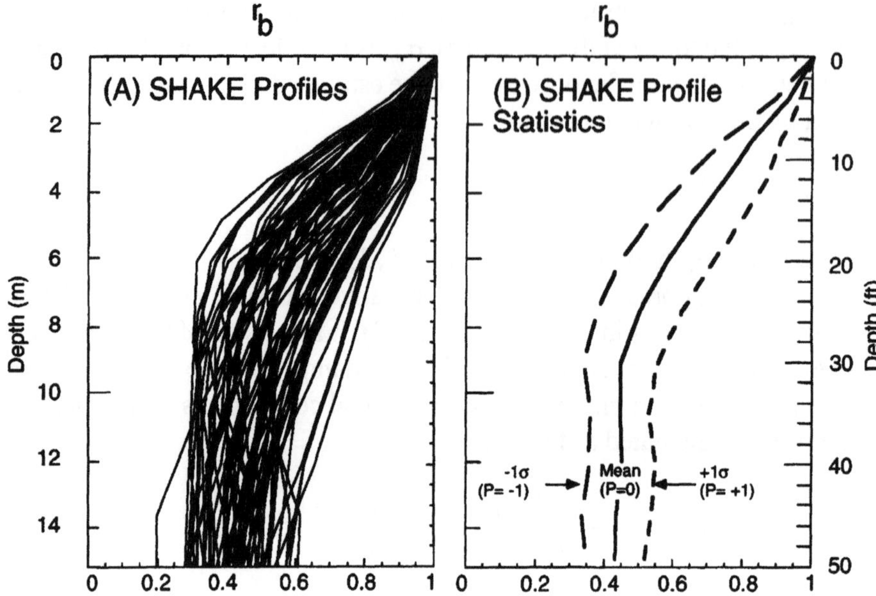

Figure 12. Arias intensity depth reduction profiles modeled using the ground response program SHAKE are presented in Figure 12a. A statistical synthesis of the SHAKE-study is presented in Figure 12b.

The r_b parameter is analagous to r_d and can be calculated from either one- or two-component horizontal Arias Intensity. The mean and $\pm 1\sigma$ responses calculated from the suite of SHAKE90 runs are plotted in Figure 12. The mean attenuation of r_b falls somewhat linearly from 1.0 to a value of 0.58 at 6 m depth, then decreases to 0.46 at 10 m. The value of r_b remains essentially constant at depths below 10 meters (30.5 ft). We calculated the Arias intensity in the soil column, I_{hb}, as the product of I_h measured from recorded seismograms in the east bay, and the reduction parameter r_b estimated from the SHAKE90 study.

$$I_{hb} = I_h \times r_b \tag{11}$$

In our study, we associate CSR and I_{hb} profiles with the liquefaction field performance data $(N_1)_{60}$, and q_{c1} from our study sites (Figure 13). Detailed discussion of the Arias intensity method for liquefaction potential assessment is presented in Kayen (1993), Kayen and Mitchell (1997), and Kayen *et al.* (1997). CSR and I_{hb} values used in the analyses of critical soil layers at the east bay study sites (Figure 1) are summarized in Table I.

The SPT boundary curve for the assessment of soil liquefaction potential based on Arias intensity is presented in detail in Kayen and Mitchell (1997) and is presented here for soils with low-plasticity fines content of less than 5 % (Figure 13a). An equivalent cyclic stress ratio based boundary for assessment of liquefaction potential was presented by Seed and de Alba (1986). The insensitivity of CSR to

Table I. Representative values of parameters used in liquefaction-susceptibility analyses for critical soil layers at east bay study sites (see Figures 1, 2, 3, and 4, for locations of boreholes. PGA, peak ground acceleration; q_{c1}, normalized cone-penetration resistance; $(N_1)_{60}$, 60%-hammer-efficiency normalized standard penetration resistance; CSR, cyclic stress Ratio; I_{hb}, Arias intensity; f_c, fines content; $(N_1)_{60}$, f_c, fines content corrected SPT-resistance)

Site	Case #	Seismogram	I_h (m/s)	PGA (g)	CSR	Depth (m)	I_{hb} (m/s)	$(N_1)_{60}$	$(N_1)_{60}, f_c$	q_{c1} (MPa)	FC (%)	Liquefaction
SF-Oakland Bay Bridge1	3	Oakland Outer Harbor	1.71	0.29	0.25	7	0.99	10	11.2	8	10	Yes
SF-Oakland Bay Bridge2	4	Oakland Outer Harbor	1.71	0.29	0.26	7.5	0.96	12	13.2	11	10	Yes
SF-Oakland Bay Bridge3	5	Oakland Outer Harbor	1.71	0.29	0.26	7	0.99			9	10	Yes
SF-Oakland Bay Bridge4	6	Oakland Outer Harbor	1.71	0.29	0.26	7	0.99			5	10	Yes
SF-Oakland Bay Bridge5	7	Oakland Outer Harbor	1.71	0.29	0.26	7	0.99	10	11.2	9	10	Yes
Treasure Island	71	Treasure Island	0.54	0.18	0.17	6	0.34	5	5.0		5	Yes
Port of Oakland, 7th St. -1	8	Oakland Outer Harbor	1.71	0.29	0.23	6.5	1.03			12	5	Yes
Port of Oakland, 7th St. -2	9	Oakland Outer Harbor	1.71	0.29	0.23	6	1	15	15.0	9	2	Yes
Port of Oakland, 7th St. -3	10	Oakland Outer Harbor	1.71	0.29	0.22	6-to-7	0.99	18	18.0	7	2	Yes
Port of Oakland, 7th St. -3	11	Oakland Outer Harbor	1.71	0.29	0.24	9	0.88	37	37.0	17	5	No
Port of Oakland, 7th St. -5	12	Oakland Outer Harbor	1.71	0.29	0.22	5	1.15			12	5	Yes
Port of Oakland, 7th St. -6	13	Oakland Outer Harbor	1.71	0.29	0.22	5.5	1.10			10	5	Yes
Bay Farm Island-Lagoon (CPT1)	14	Alameda NAS	0.8	0.27	0.23	3-to-4	0.6	16.9	16.9	7	2	Yes
Bay Farm Island-So. Loop Rd	15	Alameda NAS	0.8	0.27	0.23	3-to-4	0.6	15	15.0	12	5	Yes
Bay Farm Island-Improved Dike	16	Alameda NAS	0.8	0.27	0.22	3-to-4	0.6	38	38.0	35	5	No
Port of Richmond-2	82	City Hall, Richmond/SHAKE	0.7	0.18	0.17	6	0.44	5	12.0	1	35	Yes
Port of Richmond-3	83	City Hall, Richmond/SHAKE	0.7	0.18	0.17	6	0.44	3	10.0	1.3	35	Yes
Port of Richmond-4	84	City Hall, Richmond/SHAKE	0.7	0.18	0.17	6	0.44	3	10.0	2	35	Yes
Oakland Int'l Airport/Site 3	17	Alameda NAS	0.8	0.27	0.23	3-to-4	0.58	3	3.0	10	5	Yes
Oakland Int'l Airport/Site 4	18	Alameda NAS	0.8	0.27	0.23	3-to-4	0.58	9	9.0	5	5	Yes
Oakland Int'l Airport/Site 7	19	Alameda NAS	0.8	0.27	0.23	3-to-4	0.58	12	12.0	10	5	Yes

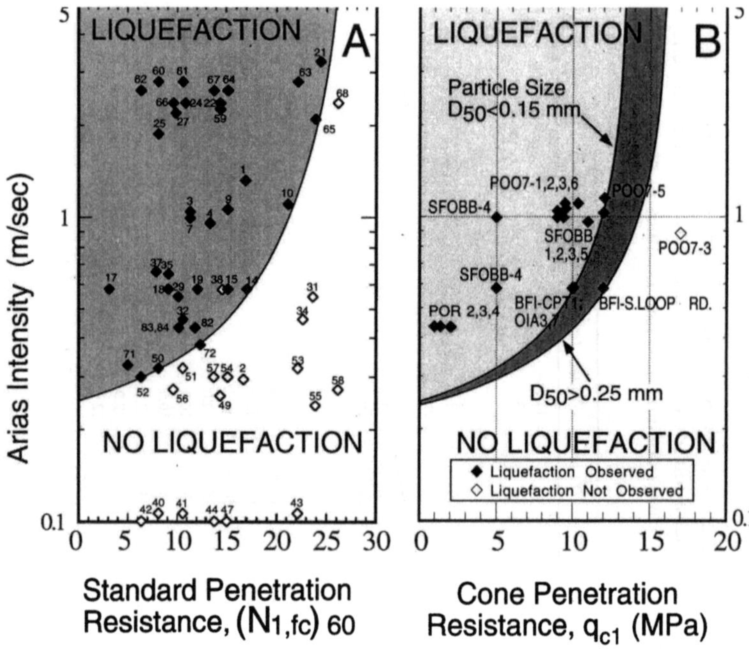

Figure 13. (a) I_{hb} versus $(N_1)_{60}$ for sites in the United States and Japan with fines content correction to equivalent 'clean sand'.; (b) I_{hb} versus q_{c1} for sites listed in Table I. Boundary curves for $D_{50} > 0.25$ mm and $D_{50} < 0.15$ mm are based on $q_c - N$ relation of Robertson and Campanella (1985).

earthquake magnitude and duration requires that magnitude correction factors be applied to CSR-values to scale for the severity of the earthquake shaking. This results in the suite of magnitude dependent boundary curves. Recent work questions the CSR boundary curves presented in Seed and de Alba (1986) as derived from the scaling factors of Seed and Idriss (1982) (Ambraseys, 1988; Arango, 1996). Arias intensity incorporates magnitude, frequency and earthquake shaking duration elements of strong motion and therefore needs no magnitude correction factors.

The I_{hb} at the liquefaction boundary in Figure 13A represents the minimum threshold seismic intensity ($I_{hb,1}$) required to induce liquefaction in soil of a given $(N_1)_{60}$ or q_{c1} value. A similar threshold in terms of CSR can be determined based on the boundaries presented by Seed and de Alba (1986). CPT-based boundary curves for CSR have been proposed by Robertson and Campanella, 1985; Seed and de Alba, 1986; Shibata and Teparaksa (1988) and Mitchell and Tseng (1990). The proposed boundaries of Robertson and Campanella (1985) and Seed and De Alba (1986) are based on q_c/N ratios rather than direct field investigations of liquefaction test sites. The method of Shiibata and Teparaksa (1988) is based on a limited data set of direct measurements at liquefaction sites for earthquakes in Japan, China, and the United States. Mitchell and Tseng (1990) presented boundary curves developed from $q_c - D_r$ (relative density)-CSR relationship at a given effective confining stress which are based on cavity expansion theory and cone chamber test results.

We compared CSR_1 and $I_{hb,1}$ with the corresponding seismic-shaking intensity function induced by the earthquake (CSR_{eq} and $I_{hb,\,eq}$) using Equations (7) through (10) to determine profiles of factor of safety against liquefaction. The Arias intensity based factor of safety $F_{I_{hb}}$ against liquefaction occurrence is defined as the ratio of the Arias intensity required to cause liquefaction to the Arias intensity imparted by the earthquake.

$$F_{I_{hb}} = \frac{I_{hb,1}}{I_{hb,\,eq}}. \tag{12}$$

This Arias intensity definition of factor of safety is analogous to that defined by Seed and his colleagues based on cyclic stress (F_{CSR}).

We converted the N profiles from Figures 5–11 into factor of safety profiles for both $F_{I_{hb}}$ and F_{CSR} in Figure 14, based in the calculated CSR and I_{hb} for the study sites. At Bay Farm Island both methods accurately predict no-liquefaction at the improved-dike and liquefaction at between 3.5 and 4.5 m at HBI-CPT1, and at Oakland International Airport both methods predict liquefaction in a zone between approximately 2 and 5 m. The profiles at the Port of Oakland for POO7-2 are in agreement and identify the zone of liquefaction between 5 and 8 m, and POO7-3 indicates thins seams of loose potentially liquefiable material at 6.5 and 7.5 m, but otherwise a profile of non-liquefaction susceptibility. Profiles of F_{CSR} and $F_{I_{hb}}$ for the San Francisco–Oakland Bay Bridge site correctly indicate that liquefaction occurred principally at a depth of between approximately 5.5 and 7.5 m at SFOBB-1 and between 6.0 and 9.0 m at SFOBB-5, based on field observations. At the Port of Richmond, the profiles accurately identify the zone of liquefaction between 4 and 5 m in the profile POR-2. At POR-Hall St., both profiles accurately indicate no liquefaction.

The CPT data used to construct the $I_{hb} - q_{c1}$ association in Figure 13b are derived almost exclusively from the Loma Prieta dataset. Thus, $F_{I_{hb}}$ profiles, when presented in a manner similar to Figure 14, are entirely consistent and in agreement with the results presented in Figure 13b where soil layers are summarized as individual I_{hb}/q_{c1} points. The new Arias intensity boundary for liquefaction asseesment, presented here, is therefore based on a limited data set from the Loma Prieta earthquake. This boundary will enhanced by new liquefaction field data that become available.

5. Conclusions

Extensive soil liquefaction occurred during the 1989 Loma Prieta earthquake in uncompacted artificial-fill deposits of the east bay from Oakland International Airport to the Port of Richmond, from 65 to 85 km from the north end of the rupture zone. We present the results of studies at five sites near the east bay

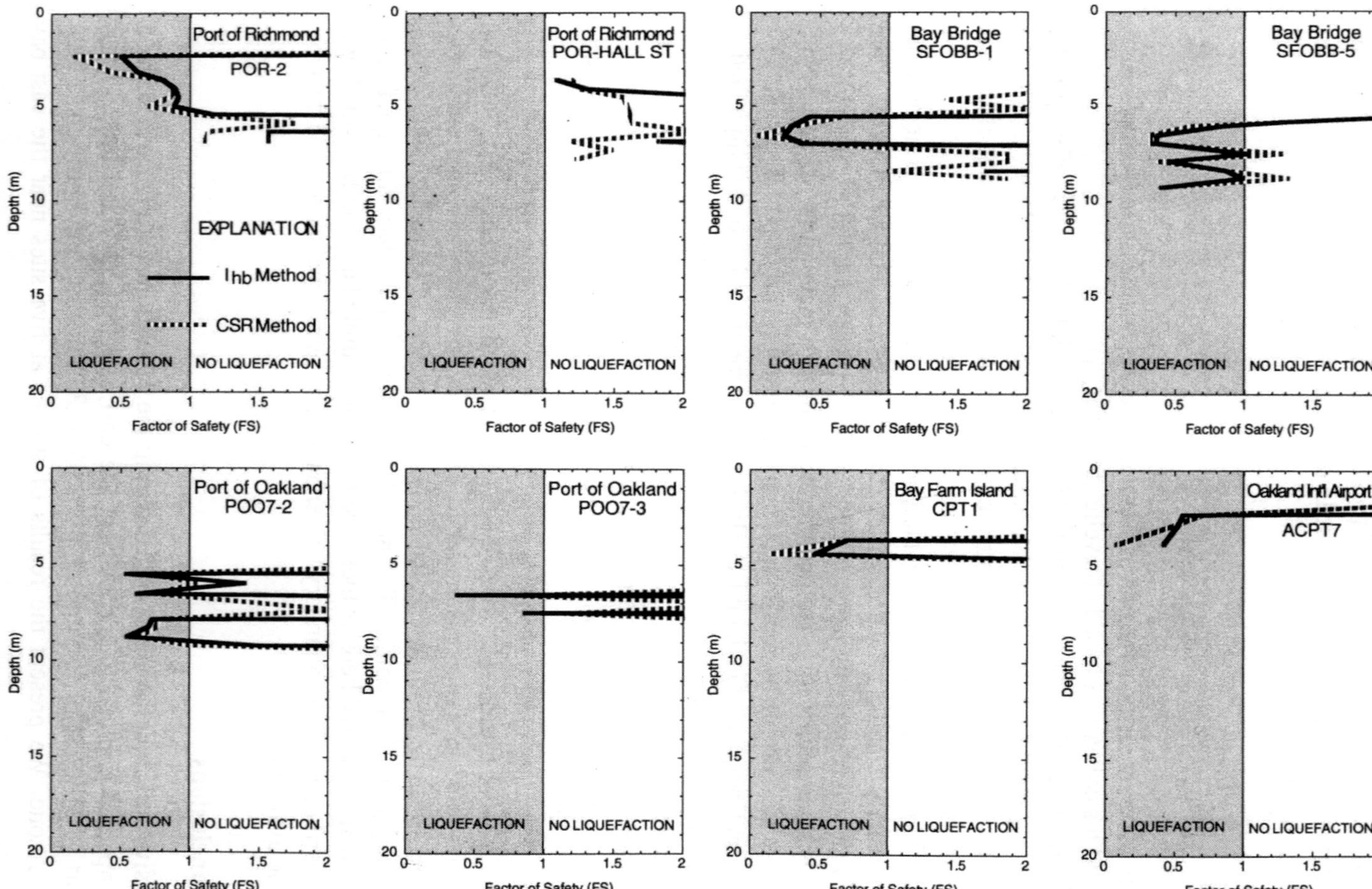

Figure 14. Comparison of the factor of safety against liquefaction (FS $\leq$ 1.0 = liquefaction; FS > 1.0 = no liquefaction) for the study sites on the east side of San Francisco Bay using the Arias intensity and the Cyclic Stress methods.

shoreline: at the Port of Richmond, San Francisco–Oakland Bay Bridge Toll Plaza, Port of Oakland's Seventh St. Marine Container Facility, Bay Farm Island and Harbor Bay Plaza, and Oakland International Airport. Typical of all these sites are extremely-low to low N and q_c values in layers of cohesionless hydraulic fill that overlie deep, primarily cohesive soil deposits. Two factors, low penetration resistance and amplification of seismic shaking by the underlying soils, combined to give the sites a relatively high liquefaction susceptibility.

A direct comparison was made of the cyclic stress and Arias intensity-based methods for assessing the potential of soil liquefaction using the SPT field test. We found that when using SPT data, the two methodologies are largely in agreement in predicting the occurrence or nonoccurrence of liquefaction in soils layers at our study sites during the earthquake. Field data from the Loma Prieta study sites allow us to construct a new Arias intensity-cone penetration resistance boundary for liquefaction assessment.

For liquefaction assessment, the advantages in using Arias intensity over peak ground acceleration (a_{max}), used in the cyclic-stress approach, are that (1) Arias intensity is derived from the acceleration records of both horizontal components of motion over the entire duration of motion, whereas a_{max} utilizes a single arbitrarily selected value; (2) Arias intensity incorporates the intensity of motions over the full range of recorded frequency, whereas, a_{max} is often associated with high-frequency motion; and (3) the breakdown of soil structure that results in liquefaction is fundamentally more dependent upon input-energy than on a single level of acceleration (Liang, and others., 1996).

References

Ambraseys, N. N.: 1988, Engineering seismology, *Earthquake Eng. Struct. Dynam.* **17**(1), 1–105.

American Society for Testing and Materials: 1984, Standard method for penetration test and split-barrel sampling of soils, *Annual Book of ASTM Standards*, Section 4, Volume 04.08, D1586-84. American Society for Testing and Materials: 1986, Standard method for quasi-static, cone and friction cone penetration test of soil, *Annual Book of ASTM Standards*, Section 4, Volume 04.08, D3441-86. American Society for Testing and Materials: 1986, Standard test method for stress wave energy measurement for dynamic penetrometer testing systems: *Annual Book of ASTM Standards*, Section 4, Volume 04.08, D4633-86, pp. 943–946.

Arango, I.: 1996, Magnitude scaling factors for soil liquefaction evaluations, *ASCE J. Geotech. Eng.* **122**(11), 929–936.

Borchedt, R .D. (ed.): 1994, The Loma Prieta, California, Earthquake of October 17, 1989-Strong Ground Motion, *U.S. Geological Survey Professional Paper 1551-A*.

Campanella, R. G. and Robertson, P. K.: 1982, State-of-the-art in in-situ testing of soils; developments since 1978, *Engineering Foundation Conference on Updating Subsurface Sampling of Soils and Rocks and Their In-Situ Testing*, Santa Barbara, Calif., Proceedings, p. 23.

Farrar, J. A.: 1991, *Field Energy Measurements of Standard Penetration Testing*, M.S. Thesis, Denver, University of Colorado, Denver, 400 pp.

Hardin, B. O. and Drnevich, V. P.: 1972, Shear modulus and damping of soils: Measurement and parameter effects, *Journal of Soil Mechanics and Foundation Division* ASCE **98**(SM6), Proc. Paper 8977.

Kayen, R. E.: 1993, *Accelerogram-Energy Approach for Prediction of Earthquake-Induced Ground Liquefaction*, PhD Dissertation, University of California, Berkeley, 289 pp.

Kayen, R. E., Mitchell, J. K., Lodge, A., Seed, R. B., Nishio, S., and Coutinho, R.: 1992, Evaluation of SPT-, CPT-, and shear wave-based methods for liquefaction potential assessment using Loma Prieta data: Proceedings of the Fourth Japan - U.S.Workshop on Earthquake Resistant Design of Lifeline Facilities and Countermeasures for Soil Liquefaction, in M. Hamada and T. D. O'Rourke (eds.), *Technical Report NCEER-92-0019*, Vol. 1, pp. 177–204.

Kayen, R. E., Mitchell, J. K., and Holzer, T. L.: 1994, Ground motion charateristics and their relation to soil liquefaction at the Wildlife Liquefaction Array, Imperial Valley, California: Proceedings of the Fifth U.S. – Japan Workshop on Earthquake Resistant Design of Lifeline Facilities and Countermeasures for Soil Liquefaction, in T. D. O'Rourke and M. Hamada (eds.), *Technical Report NCEER-94-0026*, pp. 267–283.

Kayen, R. E. and Mitchell, J. K.: 1997, Assessment of liquefaction potential during earthquakes by Arias intensity, *Am. Soc. Civil Eng.: J. Geotech. Geoenviron. Eng.* **123**(12), 1162–1174.

Kayen, R. E., Mitchell, J. K., Seed, R. B., and Nishio, S.: 1998, Soil Liquefaction along the eastern shoreline of San Francisco Bay during the Loma Prieta Earthquake, 17 October 1989, in T. L. Holzer, (eds.), *The Loma Prieta, California, Earthquake of October 17, 1989, Professional Paper II – Earthquake Effects: U.S. Geological Survey Professional Paper 1551-B*, NEHRP Report to Congress.

Mitchell, J. K. and Tseng, D.-J.: 1990, Assessment of liquefaction potential by cone penetration resistance, in J. M. Duncan (ed.), *H. Bolton Seed Memorial Symposium*, Vol. 2, Berkeley, Calif., BiTech, pp. 335–350.

Mitchell, J .K., Lodge, A. L., Coutinho, R. Q., Kayen, R. E., Seed, R. B., Nishio, S, and Stokoe, K. H. II: 1994, In situ test results from four Loma Prieta earthquake liquefaction sites: SPT, CPT, DMT and Shear Wave Velocity: Berkeley, University of California Earthquake Engineering Research Center Report UCB/EERC-94/04, 179 pp.

Robertson, P.K., 1986. In situ testing and its application to foundation engineering: *Can. Geotech. J.* **23**(4), 573–594.

Robertson, P. K.: 1990, Soil classification using the cone penetration test, *Can. Geotech. J.* **27**(1), 151–158.

Robertson, P. K. and Campanella, R. G.: 1985, Liquefaction of sands using the CPT, *Am. Soc. Civil Eng. Proc., Geotech. Div. J.* **111**(GT3), 384–403.

Robertson, P. K., Campanella, R. G., Gillespie, Don, and Rice, A.: 1986, Seismic CPT to measure in situ shear wave velocity, *Journal of Geotechnical Engineering* **112**(8), 781–803.

Schnabel, P. B., Lysmer, John, and Seed, H. B.: 1972, SHAKE – a computer program for earthquake response analysis of horizontally layered sites, *Report UCB/EERC-72/12*, Earthquake Engineering Research Center, University of California, Berkeley.

Seed, H. B. and de Alba, Pedro: 1986, Use of SPT and CPT tests for evaluating the liquefaction resistance of sands, in S. P. Clemence (ed.), *Use of In-Situ Tests in Geotechnical Engineering*, American Society of Civil Engineers, New York, pp. 281–302.

Seed, H. B. and Idriss, I. M.: 1971, Simplified procedure for evaluating soil liquefaction potential, *Am. Soc. Civil Eng. Proc., Soil Mech. Found. Div. J.* **97**(SM9), 1249–1273.

Seed, H. B. and Idriss, I. M.: 1982, *Ground Motions and Soil Liquefaction during Earthquakes*, Earthquake Engineering Research Institute, University of California, Berkeley, 139 pp.

Seed, H. B., Idriss, I. M., and Arango, Ignacio: 1983, Evaluation of liquefaction potential using field performance data, *J. Geotech. Eng. ASCE* **109**(3), 458–482.

Seed, H. B., Tokimatsu, Kohji, Harder, L. H., and Chung, R. M.: 1984, The influence of SPT procedures in soil liquefaction evaluations, *Report UCB/EERC-90/05*, Earthquake Engineering Research Center, University of California, Berkeley, 50 pp.

Seed, R. B., Dickenson, S. E, Reimer, M. F., Bray, J. D., Sitar, Nicholas, Mitchell, J. K., Idriss, I. M., Kayen, R. E., Kropp, Alan, Harder, L. F., Jr., and Power, M. S., 1990, Preliminary report on the principal geotechnical aspects of the October 17, 1989 Loma Prieta earthquake, *Report UCB/EERC-90/05*, Earthquake Engineering Research Center, University of California, Berkeley, 137 pp.

Shakal, A. F., Huang, M. J., Reichle, Michael, Ventura, C. E., Cao, Tianging, Sherbourne, R. W., Savage, M. K., Darragh, R. B., and Petersen, C. P.: 1989, CSMIP strong-motion records from the

Santa Cruz Mountains (Loma Prieta), California earthquake of 17 October 1989, *Report OSMS 89-06*, Office of Strong Motion Studies, California Division of Mines and Geology, 196 pp.

Shibata, Toru and Teparaksa, Wanchai: 1988, Evaluation of liquefaction potentials of soils using cone penetration tests, *Soils and Foundations*, **28**(2), 49–60.

Spudich, P. (ed.): 1996, *The Loma Prieta*, California, Earthquake of October 17, 1989 – Main-shock characteristics, *U.S. Geological Survey Professional Paper 1550-A*, pp. 297.

Wroth, C. P.: 1984, Interpretation of in situ soil test, 24th Rankine Lecture, *Geotechnique* **34**, 449–489.

Natural Hazards **16:** 267–285, 1997.

The Behaviour of a Man-Made Island During the Great Hanshin Earthquake, Japan

YASUO TANAKA
Research Centre for Urban Safety and Security, Kobe University, Nada, Kobe, Japan 657

(Received: 27 February 1997; in final form: 20 October 1997)

Abstract. This paper describes the behaviour of Port Island in Kobe city during and after the Great Hanshin Earthquake of 17 January 1995. The island is near the city centre and was constructed on the soft seabed in Kobe Port. A vertical array of four seismometers and pore water pressure sensors recorded the response of the ground during and after the earthquake. This study shows how the stiffness of the ground dropped and recovered during and after the earthquake.

Key words: man-made island, seismic response, seabed ground, soil dynamics, pore water pressure, liquefaction.

In the early morning of 17 January 1995, the Great Hanshin Earthquake struck the cities of Kobe, Nishinomiya and Ashiya, located along the north coast of Osaka Bay, for about 15 sec. The number of casualties reached over 5500 and damage to the cities was extensive. Kobe City, with a population of 1.5 million, had the largest international port facilities in Japan. The main cause of damage to the harbour was the collapse of the shore line on reclaimed land, including Port Island and Rokko Island which are man-made islands built on the seabed near the city. These near-shore man-made islands were shaken so strongly that extensive liquefaction of fill and complete destruction of sea-walls along its perimeter resulted. The epicenter of the earthquake is located at a depth of 14 km and about 25 km west of Kobe City (Figure 1). The magnitude of the earthquake was 7.2 on the Richter scale. It occurred at 5:46 local time on 17 January 1995. Figure 1 shows the locations of aftershocks recorded between 17 and 24 January 1995. The distribution of aftershocks suggests an alignment north-east to south-west parallel to the Rokko Mountains that rise behind Kobe City (Figure 2). Most of the earthquake damage and the casualties were found south of the Rokko Mountains along a very narrow strip as shown in the figure. The shaded zone in Figure 2 designates intensity 7 area on the Japanese Intensity Scale of seven degrees. It is defined by the high percentage (i.e., 30% or more) of total collapses of houses and buildings. The definition of Level 7 does not correspond to ground acceleration, but reflects the high intensity of ground motion and earthquake damage. It may be noted that the trend of Intensity 7 area does not match the north-east direction of aftershocks. Geophysical explorations and geological studies are currently in progress to examine the location of active faults and to explain the reason for this difference.

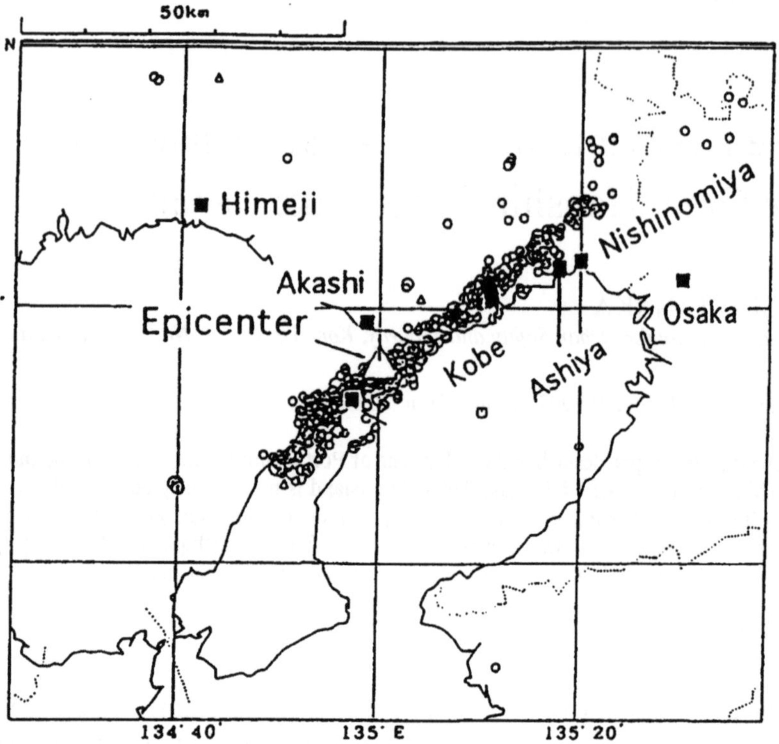

Figure 1. Location of aftershocks of the Great Hanshin earthquake.

1. Subsurface Conditions

The near-surface geology of Osaka Bay consists essentially of alternating layers of sand and clay to a considerable depth. The alternation of soil strata is a result of past global climatic changes associated with glacial advances and retreats. Figure 3 shows a geological section through Osaka Bay from velocity logging. The depth of sediments is as deep as 2000 to 3000 m, as reported by Iwasaki *et al.* (1994). The figure also shows a fault system which may be related to the current earthquake event.

The seabed stratigraphy at Kobe Port island (Figure 4) consists of a superficial layer of soft compressive marine clay of Holocene age, locally called Ma13 clay, which is underlain by a thin, relatively loose Holocene sand. In this stratum, pore water pressure built up during the earthquake as will be described later. Beneath these Holocene layers, there is a very thick layer of sand and gravel (Dg1 in Figure 4), of late Pleistocene age. Below the sand and gravel layer (Dg1), we find a layer of Upper Pleistocene medium to stiff marine clay, locally known as Ma12 clay, which is underlain by another layer of sand and gravel (Dg2) that in turn overlies yet another Upper Pleistocene stiff clay stratum, Ma11. As can be seen from Figure 4, each layer gradually increases its depth toward offshore as shown by Tanaka (1996). A recent geophysical survey shows that the bed rock is located

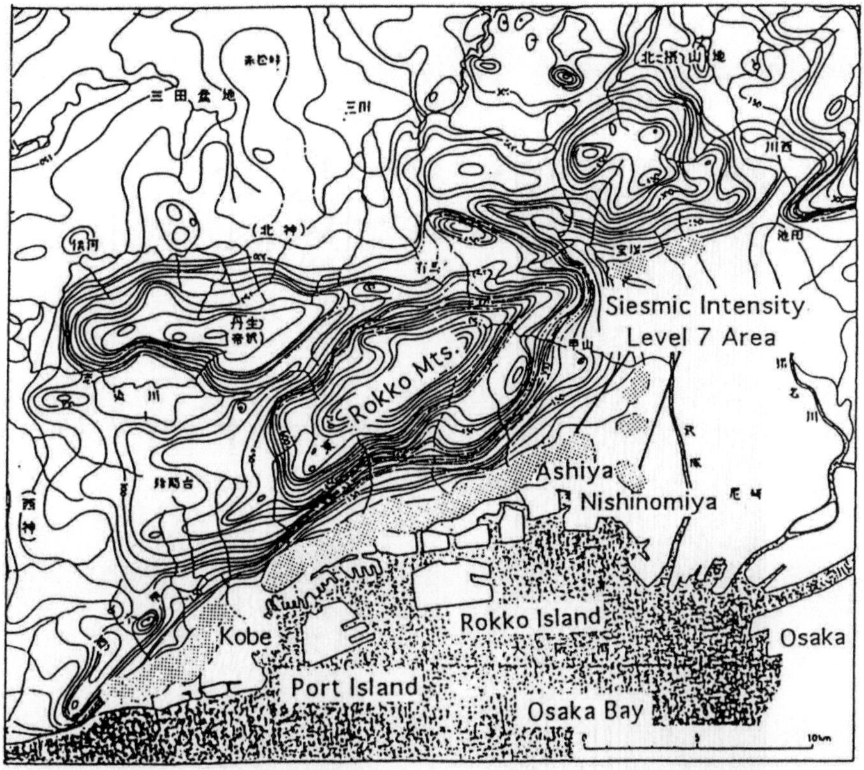

Figure 2. Topography of the Kobe area and zones of strong seismic intensity.

about 2000 m below Port Island. The water depth in Kobe Port varies between about 10 to 15 m and the elevation of fill is usually about 5 m above sea level. The first phase of construction of Port Island used granular fill, mainly of decomposed granite, emplaced on top of the soft seabed; the second phase used crushed mud stone and clay stone. At Rokko Island, which was built after Phase-1 of Port Island, the decomposed granite was used in the north part of the island and the rest of the island was built with crushed mud stone and clay stone on top of the soft marine clay.

2. Seismic Response of Reclaimed Ground

Little is known on the seismic response of man-made islands on top of a soft seabed under a strong earthquake motion. Fortunately, the geotechnical engineers at the Development Bureau of Kobe City monitored the seismic response of Port Island to the Great Hanshin Earthquake (Figure 4). The seismic record consists of digital data sampled at 1/100 second intervals spanning over 6 min. Records of horizontal acceleration for the north-south component are shown in Figures 5(a)–(d).

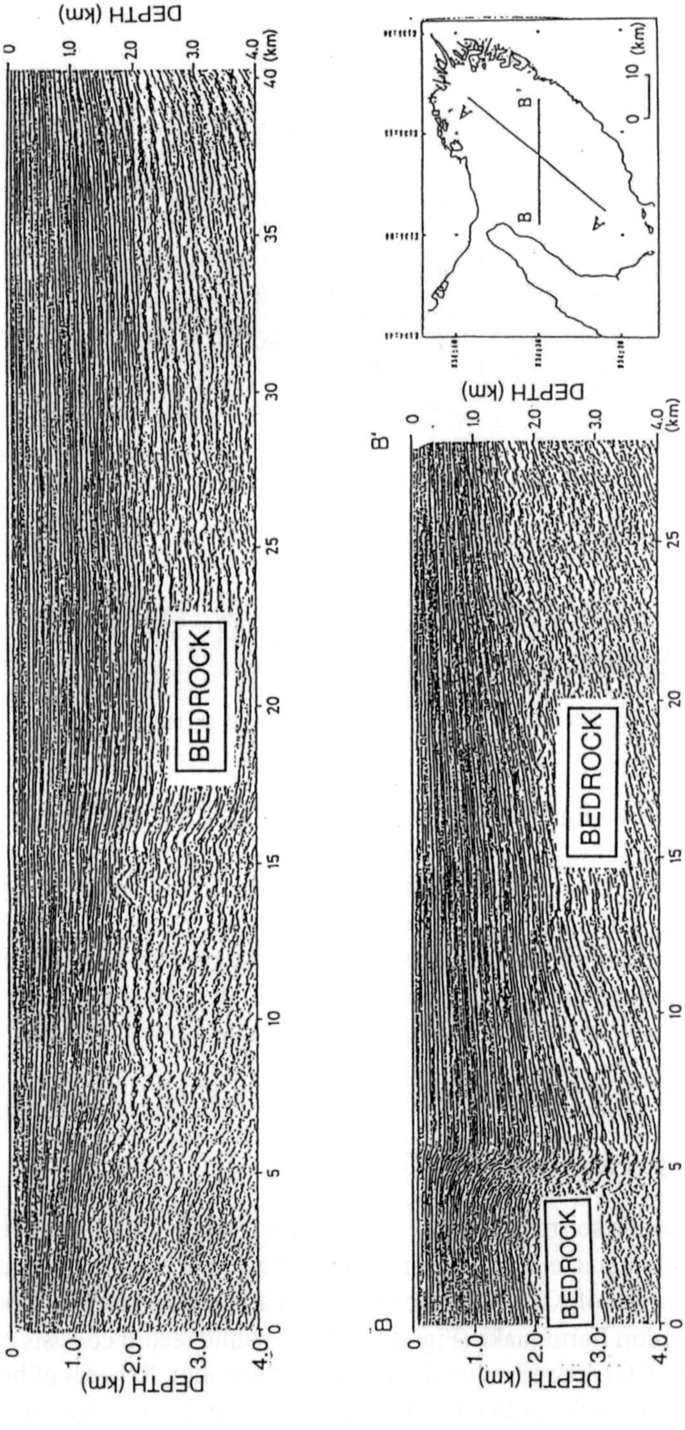

Figure 3. Geophysical sections of the Osaka basin, (Iwasaki *et al.* (1994)).

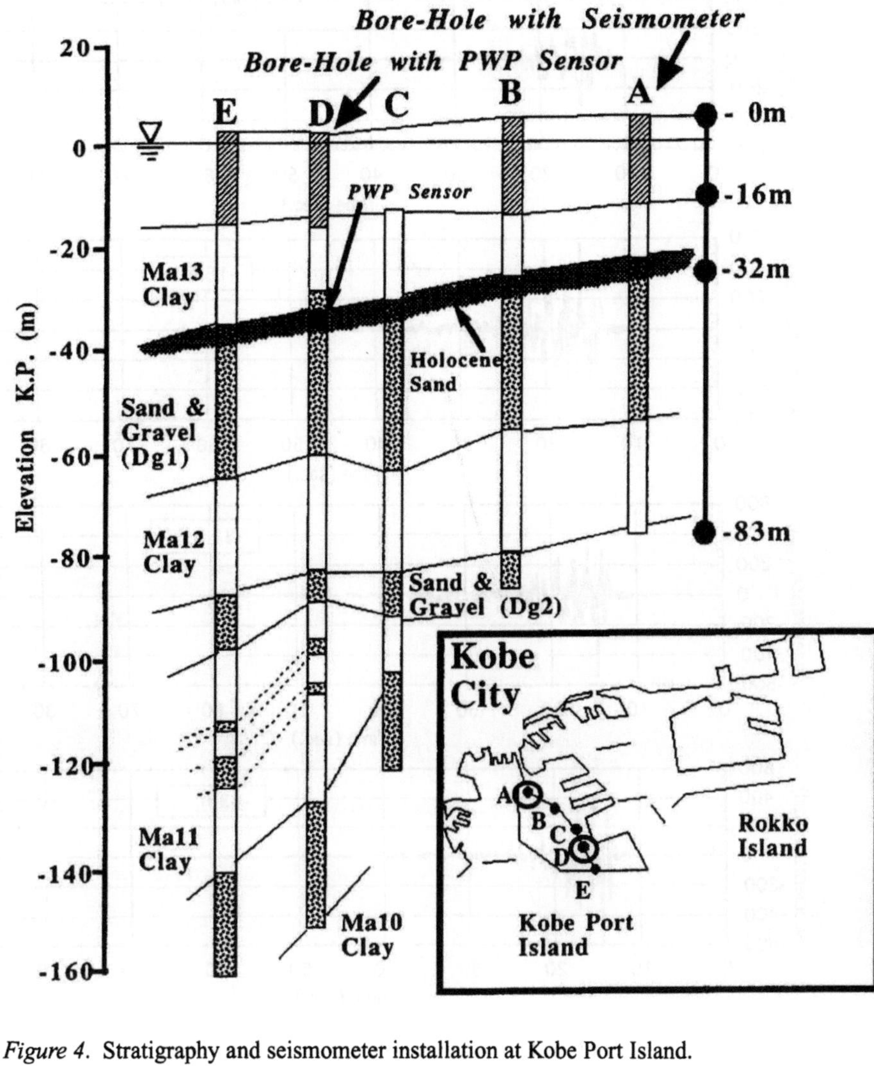

Figure 4. Stratigraphy and seismometer installation at Kobe Port Island.

Figure 5(a) shows that the main shock lasted only 15 to 20 sec, and the maximum acceleration reached 680 gals. A closer inspection reveals that the acceleration at the bottom of the hole did not decrease up to the Ma13 clay layer but the frequency response changed across layers Ma12 and Ma13. A very drastic change in the seismic record is seen at the surface. The damping of seismic motion at the surface is clear as the seismic record becomes rather smooth. In order to present more clearly the damping at the surface, the seismic records at the beginning of the quake were enlarged (Figure 6). Liquefaction of fill material occurred after one or two strong cycles. The dynamic response at the surface was drastically damped. Figure 6 can be used to calculate the seismic shear wave velocities between sensors. The results are 232.5, 84.2, and 34.2 m s^{-1} for depth ranges of 83–32, 32–16,

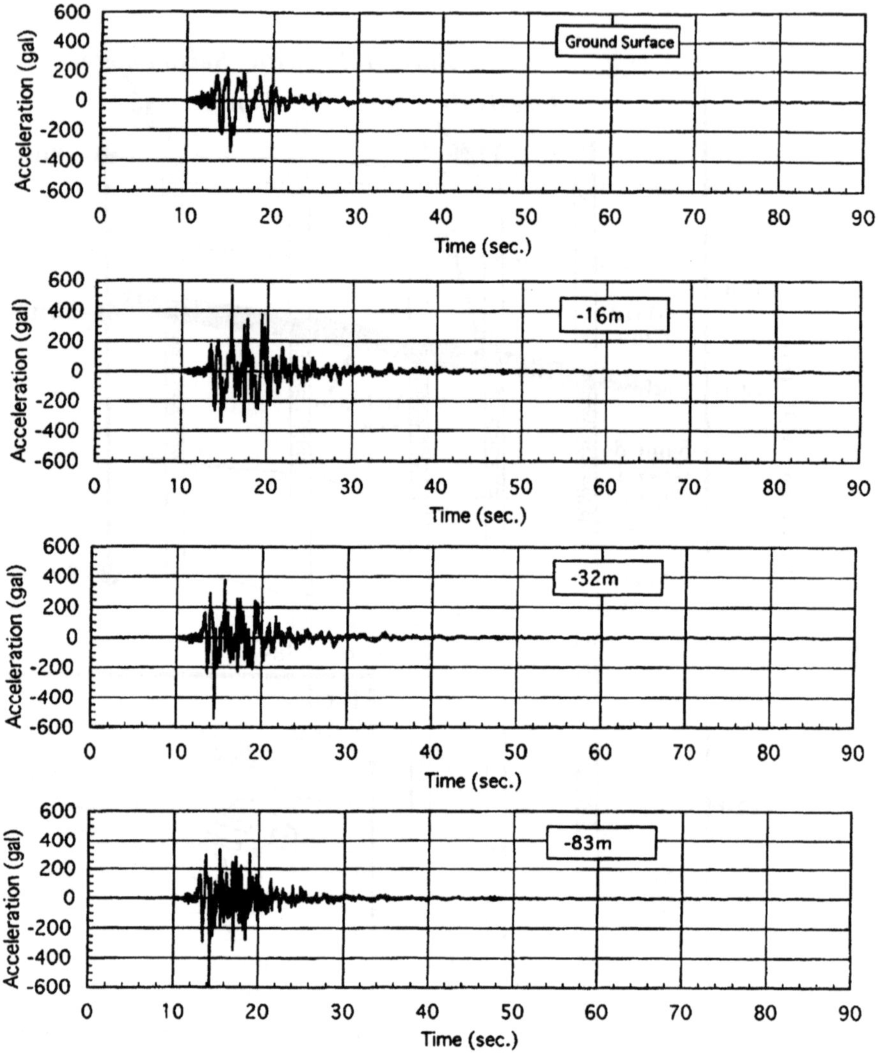

Figure 5a. Seismic records at Port Island (north-south, 0–90 sec).

and 16–0 m. The low shear wave velocity in the range of 16–0 m indicates the liquefaction of fill, and the low velocity of 84.2 m s^{-1} at the depths of 32 to 16 m indicates some softening of soil strata at that depth.

Figures 7–9 show the examples of other seismic events recorded in east-west direction. The seismic sensor for north-south component was damaged shortly after the main shock and therefore the values of V_s were computed based on data in the east-west direction. The data from events which occurred about two months before the Great Hanshin Earthquake show that the amplitudes are large, as expected for soft ground at the surface. On the other hand, the event at 5:53 (i.e., seven minutes after the main shock) shows still a drastic damping at the surface, while the data taken at 8:58 (i.e., about 3 hours after the main shock) shows already

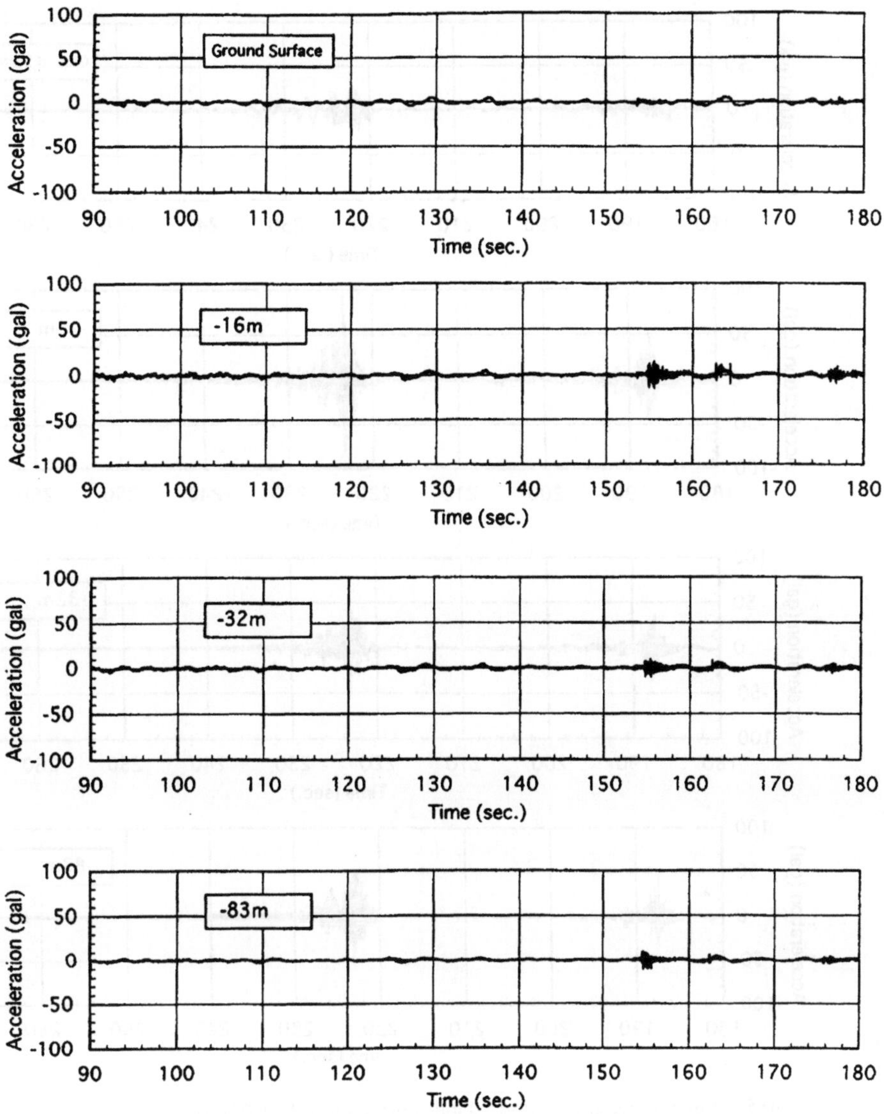

Figure 5b. Seismic records at Port Island (north-south, 90–180 sec).

some amplification at the surface. These changes of shear wave velocity with time represent stiffness changes of the ground before and after the main shock of 17 January 1995. Using these data, the shear velocities between the four observation points were tabulated in Table I. At least four to six peaks of the signal were compared and average values were obtained.

Figure 10 compares the values of V_s before and after the main shock, while Figure 11 presents the increase of V_s with time after the main shock. These data suggest that the recovery of ground stiffness occurred fairly rapidly, within a day, but some increase of V_s with time may still be lasting. The increase of ground

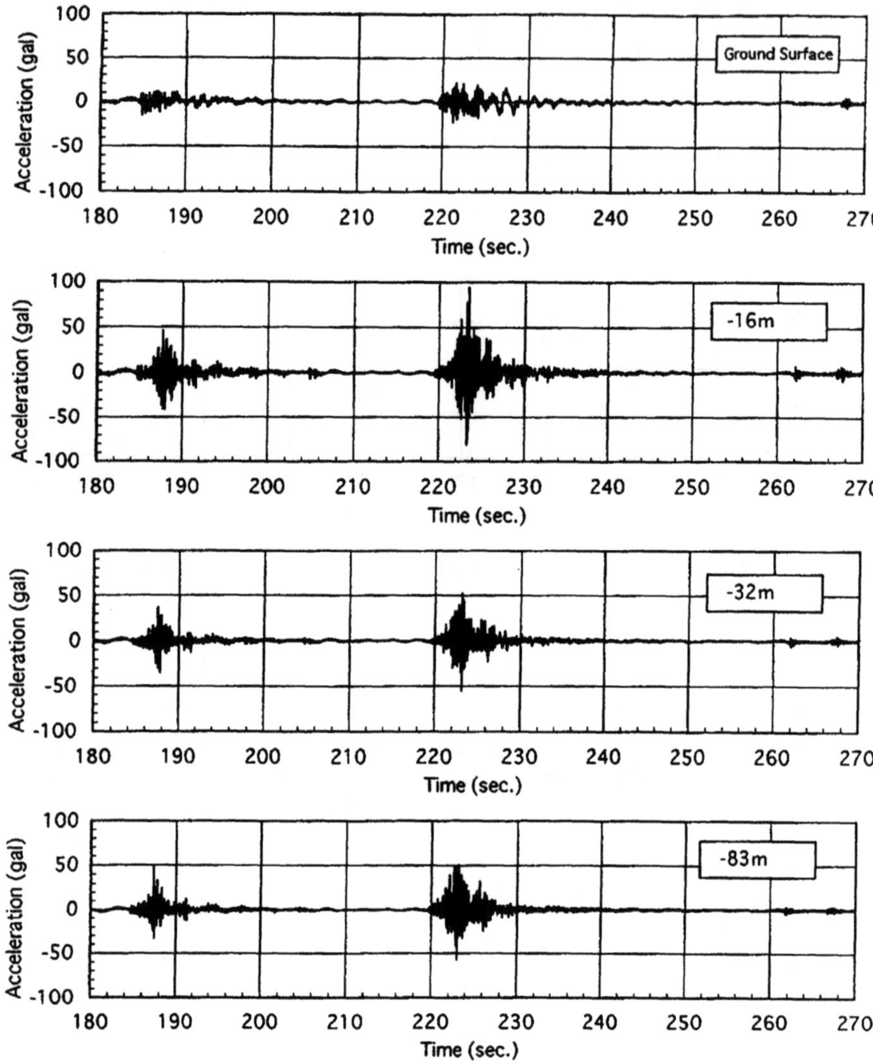

Figure 5c. Seismic records at Port Island (north-south, 180–270 sec).

stiffness is probably caused by the increase of effective stress, which may be due to the dissipation of excess pore water pressure generated by the cyclic loading during the earthquake.

3. Pore Water Pressure Measurements

Measurements of pore water pressure in the Holocene sand strata beneath the soft marine clay were made at one bore hole (D in Figure 4) located in Phase-2 of Port Island. The depth of installation was 38.5 m below the surface, and the measurement was made every 6 hr. Fortunately, the first reading on the day of earthquake was taken at 6:00 a.m., 14 min after the quake. The record of pore water pressure with

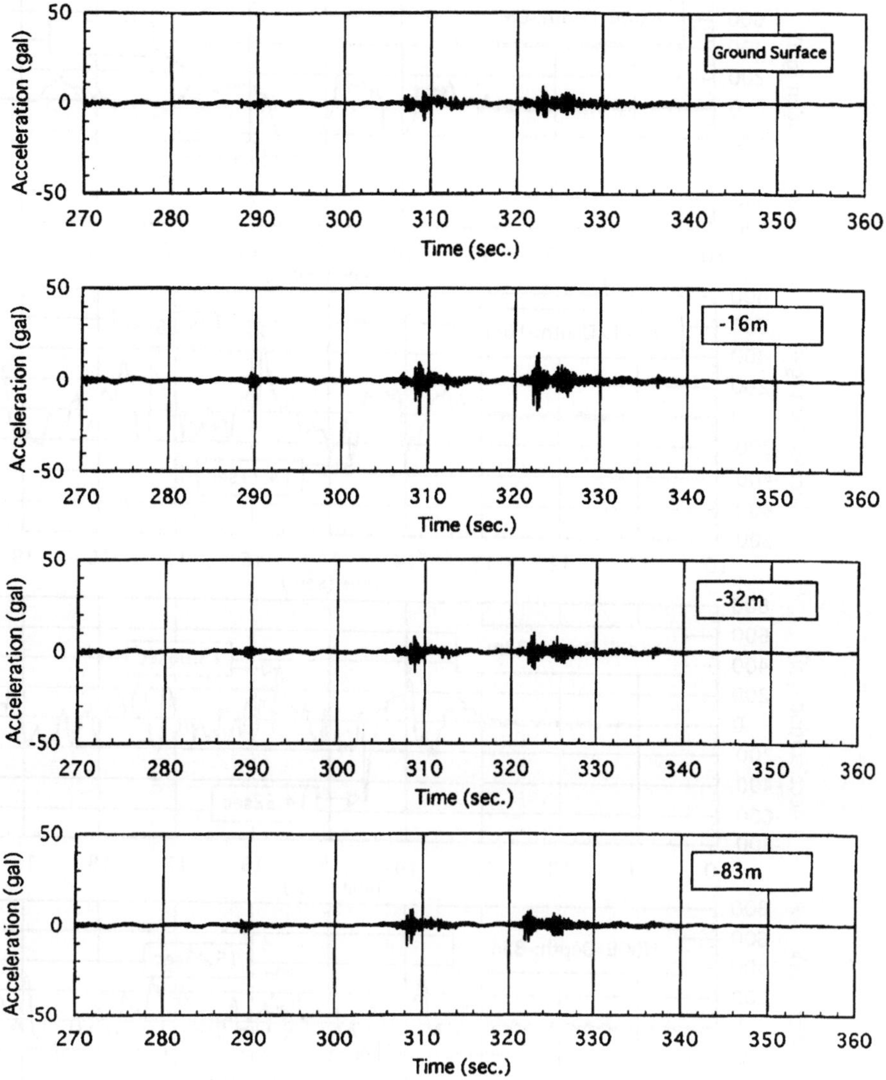

Figure 5d. Seismic records at Port Island (north-south, 270–360 sec).

time is shown in Figure 12 using arithmetic time scale. The pore water pressure rapidly decreases with time. When the same data is replotted on a logarithmic time scale (Figure 13), the pore water pressure versus time relationship is very similar to the consolidation curve of a clay.

Figure 13 indicates that there is a maximum value of excess pore water pressure build-up. This maximum is estimated at about 190 kPa, which is the water pressure above the hydrostatic pressure. Since the measurement is made at the depth of 38.5m, this excess water pressure equals to about half the effective overburden pressure. This reduction of effective stress showed have induced much lower soil stiffness during seismic motion even though the soil has not reached complete

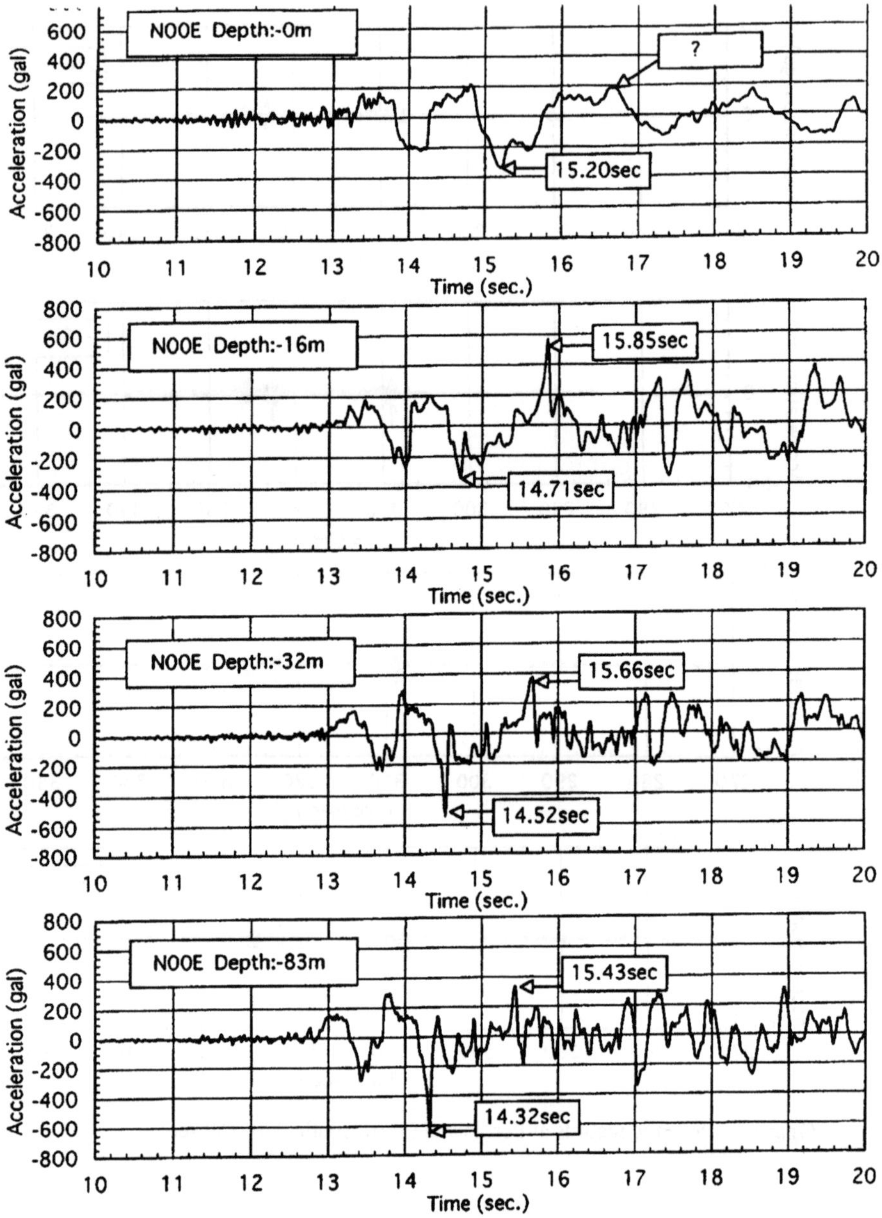

Figure 6. Initial part of seismic records at Port Island (north-south, 0–20 sec).

liquefaction. Note that it takes some time (about 3 months for this particular case) to dissipate the excess pore water pressure completely. It is surprising to observe such soil behaviour at this depth during the earthquake, and more studies are needed to improve our prediction capability of the dynamic behaviour of reclaimed land and subsurface ground during a strong earthquake.

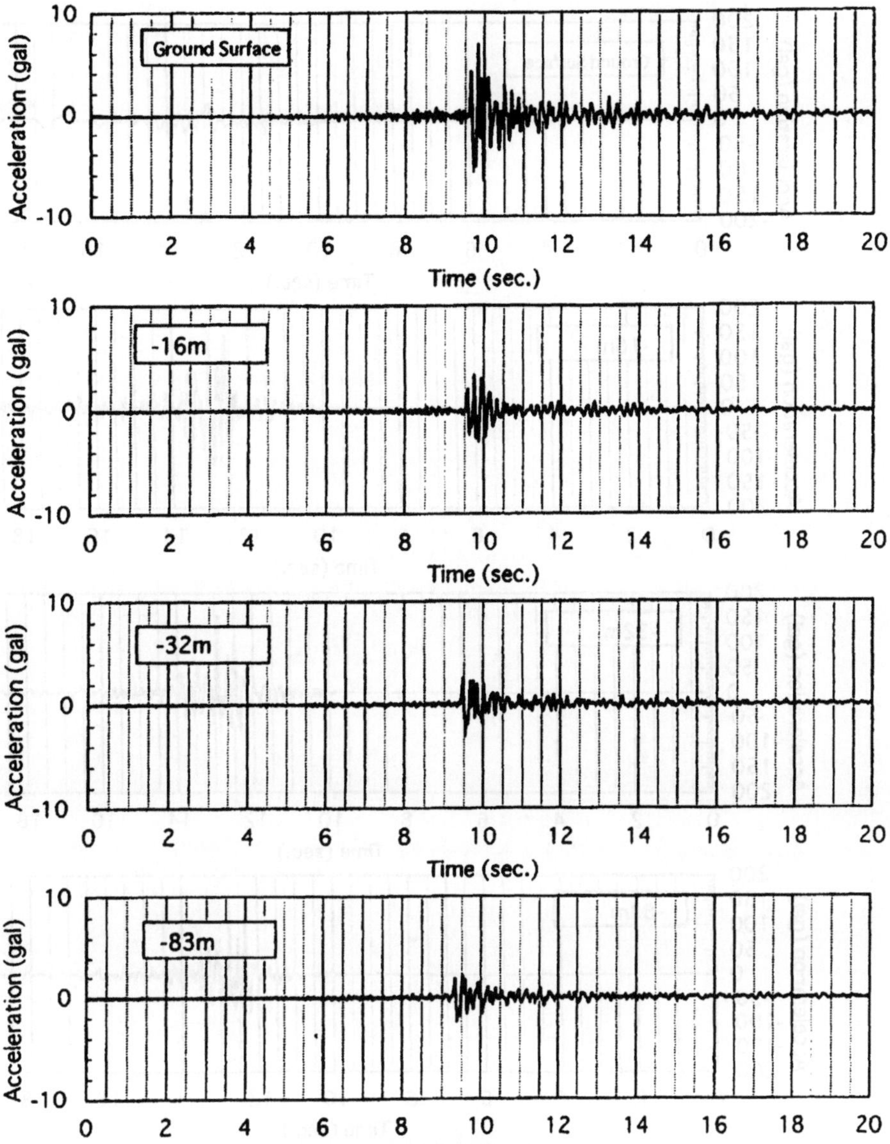

Figure 7. Seismic event prior to the Great Hanshin earthquake (east-west, 10 November 1994).

4. Discussion

Figures 12 and 13 show that the shear wave velocities in different soil layers increase with time after the main shock. The measured pore water pressure reflects the increase of effective stress in the soil layers, as the excess pore water pressure dissipates with time in a very similar manner as in consolidation. Thus the increase of shear wave velocity with time may be closely correlated with the dissipation of pore water pressure or the increase of effective stress. Figure 14 shows the time

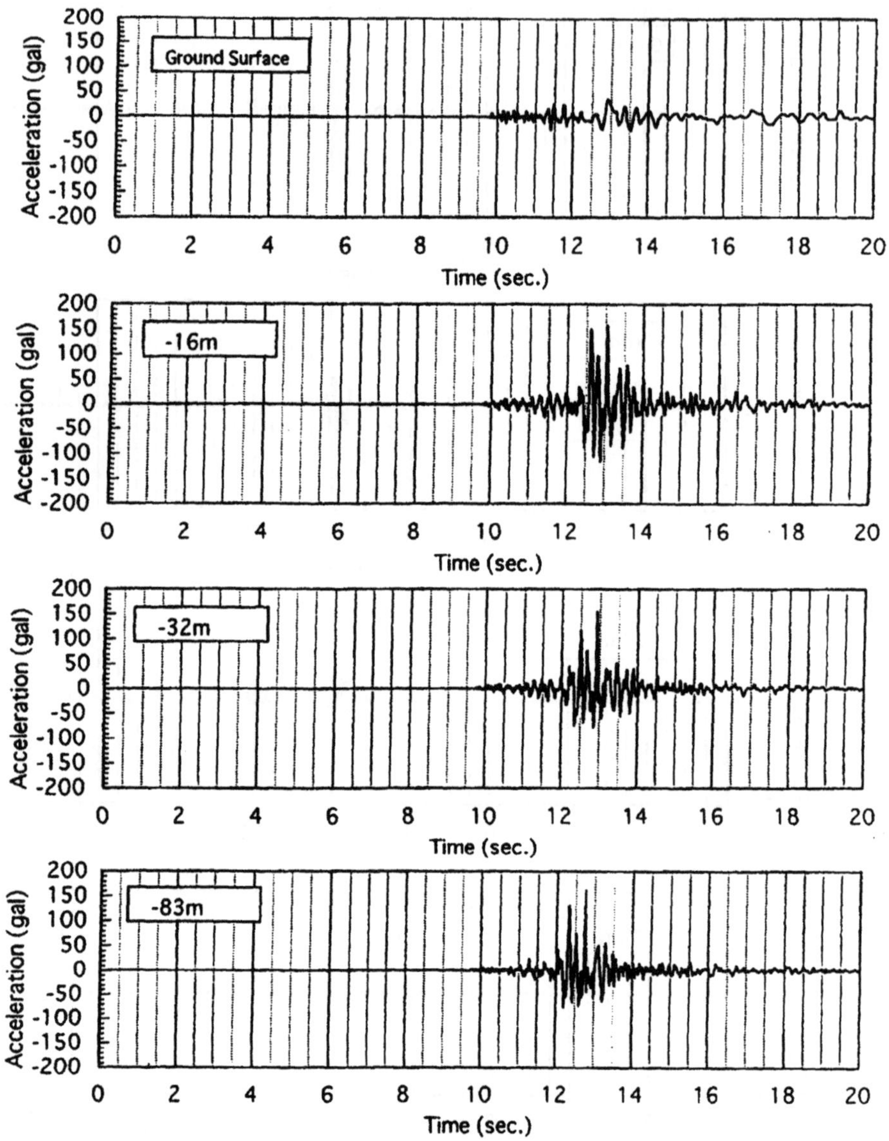

Figure 8. Aftershock of the Great Hanshin earthquake (east-west, 17 January 1995, 5: 53).

increase of V_s for different depths on a logarithmic time scale. The increase of V_s is very fast for the depth of 0–16 m, and V_s almost recovers to the value prior to the earthquake after about 10 days. The increase of V_s is also fast at the depths of 32 to 83 m. On the contrary, the increase of V_s at the depths of 16–32 m is much slower, perhaps because of the slow pore pressure dissipation in the Ma13 clay layer.

The change of V_s during the earthquake may be used to estimate the degree of effective stress change which would have been caused by the excess pore pressure

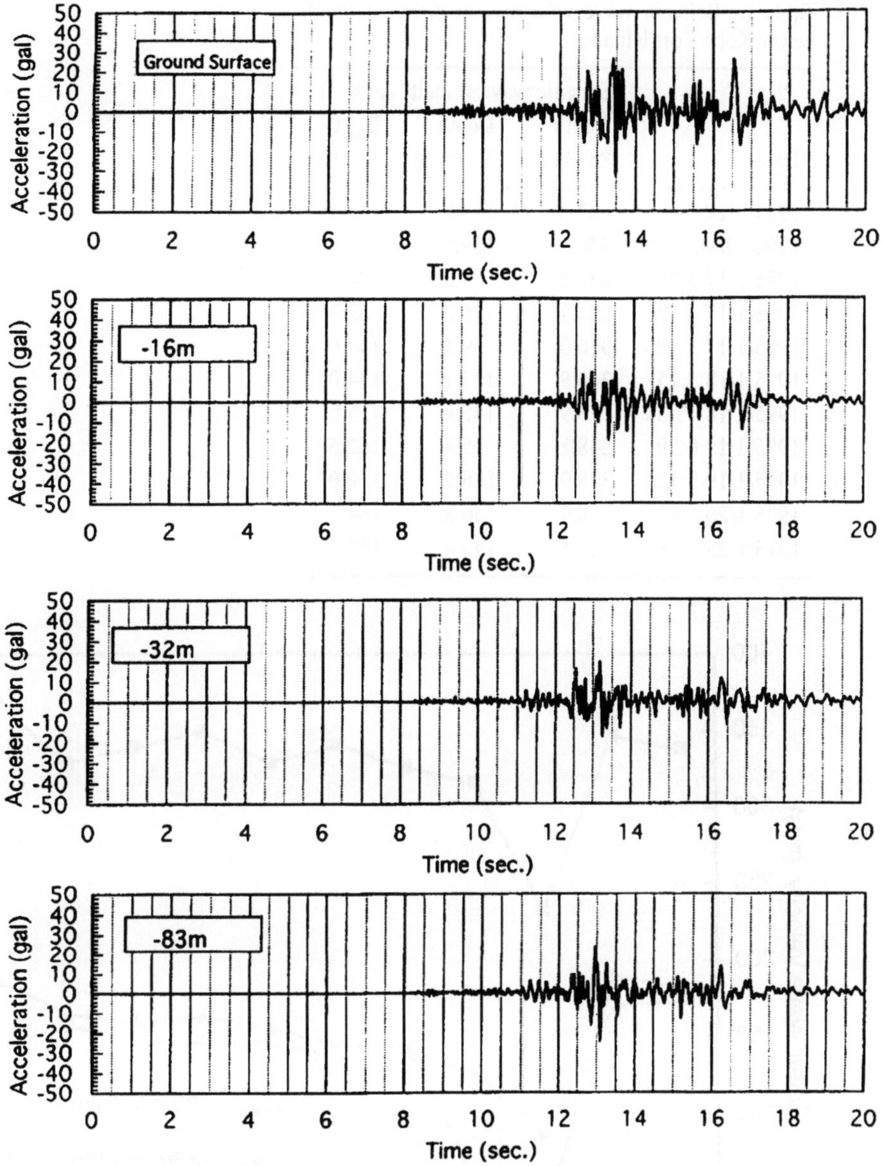

Figure 9. Aftershock of the Great Hanshin earthquake (east-west, 17 January 1995, 8:58).

build-up during earthquake loading. The following equations estimate the shear modulus, G, of a soil based either on the V_s value or on the effective stress.

$$G = \rho V_s^2 \text{ and } G = f(e)\sigma^{0.5},$$

where ρ is the bulk density of soil, and σ is the effective stress of soil. $f(e)$ is an experimentally determined parameter which is a function of the void ratio.

The former equation is usually used to determine the shear modulus from field velocity logging test, and the latter is often used to estimate the G value based

Table I. Seismic wave velocities at different depths below Kobe Port Island

	Seismic wave velocity (m s^{-1})		
Date	Depth 83–32 m	Depth 32–16 m	Depth 16–0 m
1994/6/28	337.8	173.9	182.6
1994/11/10	346.5	169.6	189.6
1995/1/17 5:46	232.5	84.2	34.2
1995/1/17 5:52	315.7	126.3	53.3
1995/1/17 8:58	326.0	139.1	144.0
1995/1/18 5:25	340.9	137.1	144.0
1995/1/18 13:34	328.9	145.4	155.5
1995/1/19 0:59	348.8	145.4	162.5
1995/1/19 5:12	338.9	148.8	160.0
1995/1/26 1:01	328.9	156.8	164.7
1995/1/28 8:06	367.6	145.4	186.6

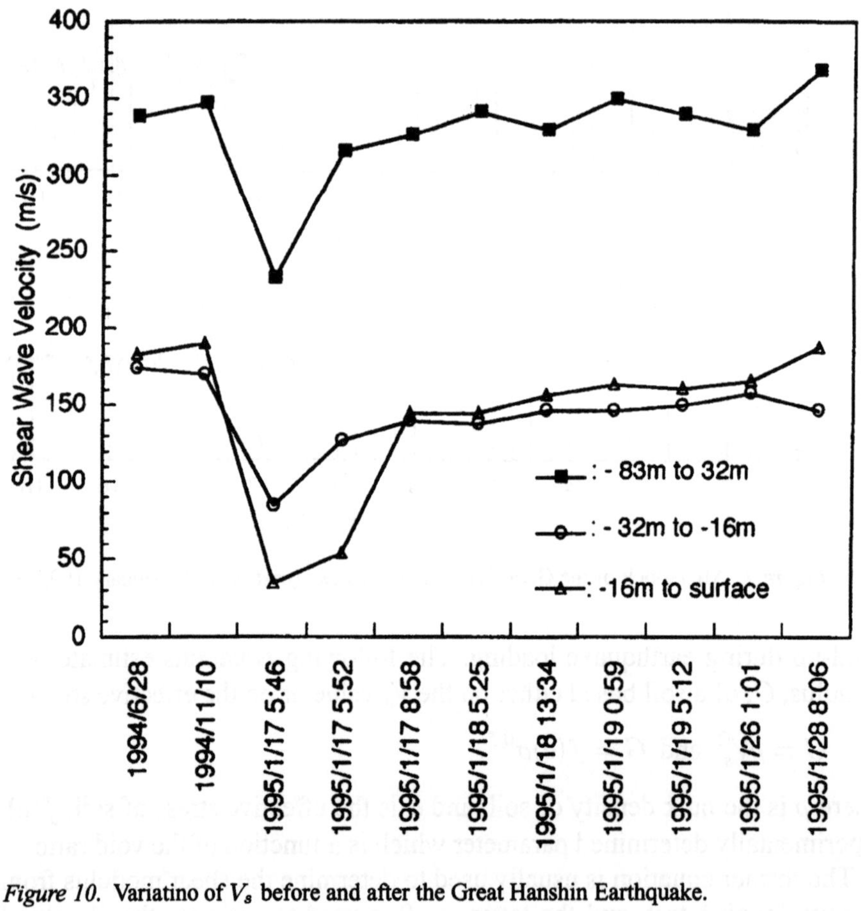

Figure 10. Variatino of V_s before and after the Great Hanshin Earthquake.

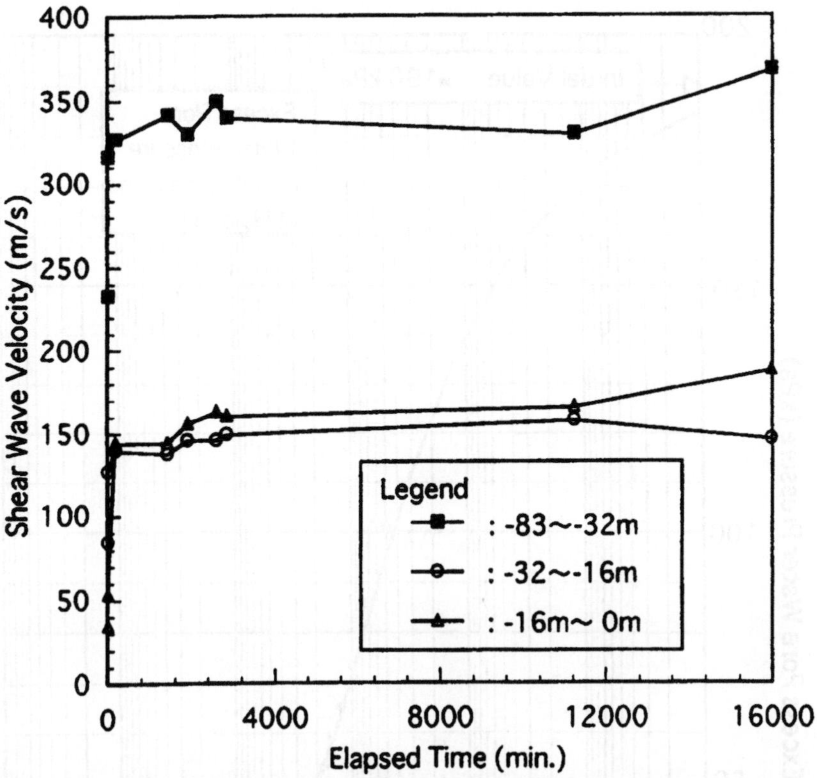

Figure 11. Increase of V_s with time after the Great Hanshin Earthquake.

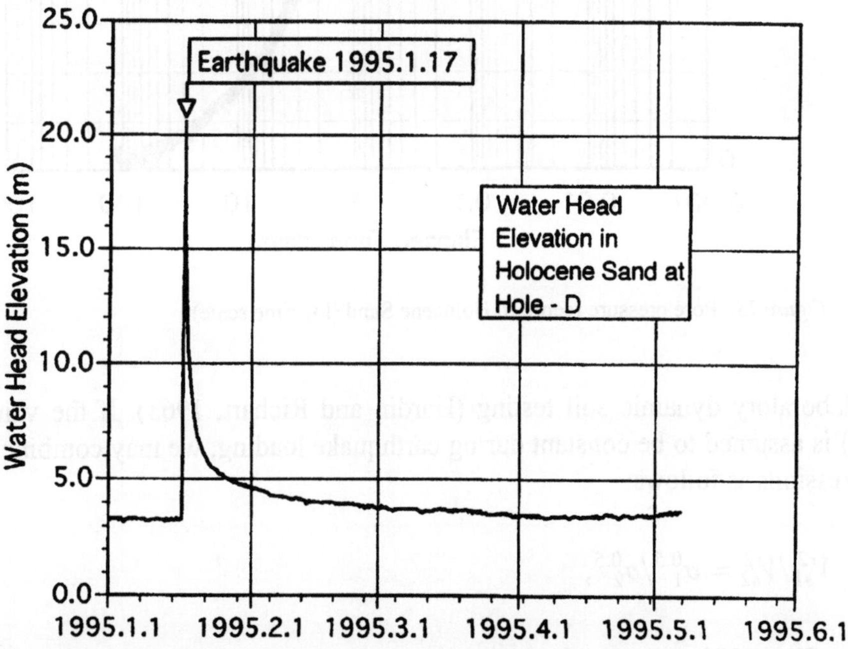

Figure 12. Pore pressure record in Holocene Sand (arithmetic time scale).

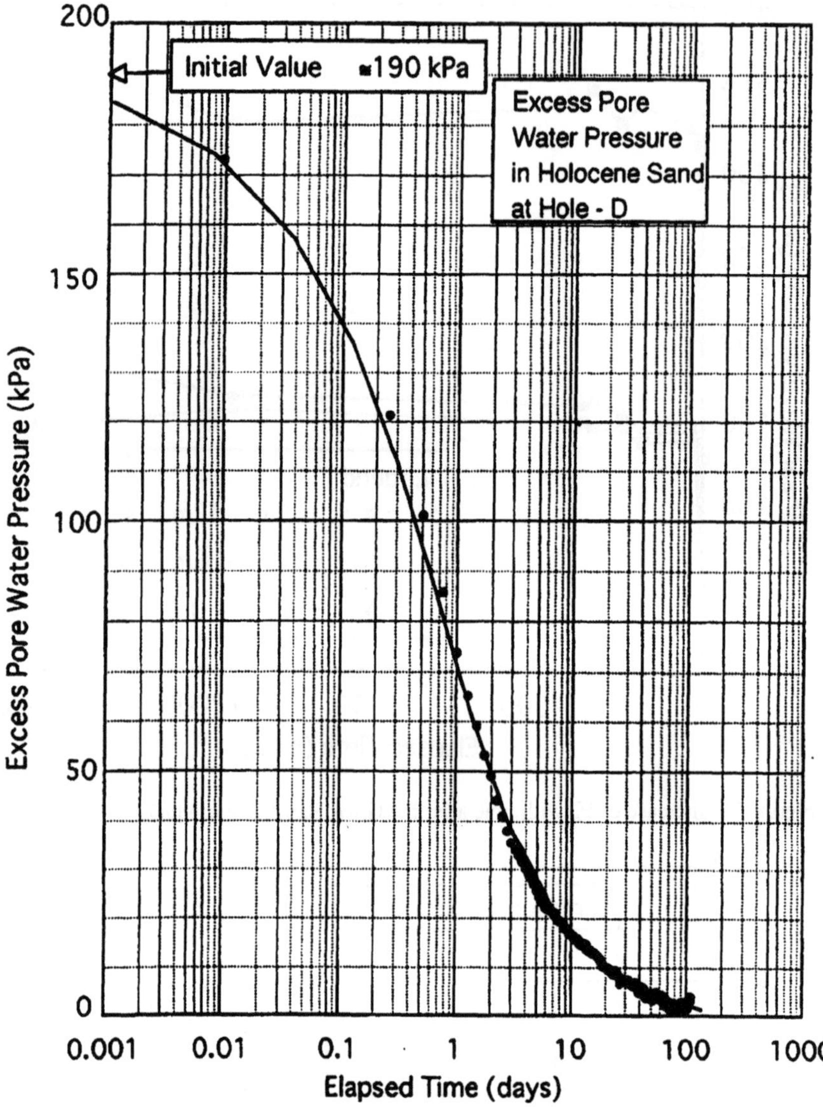

Figure 13. Pore pressure record in Holocene Sand (log time scale).

on laboratory dynamic soil testing (Hardin and Richart, 1963). If the value of $f(e)$ is assumed to be constant during earthquake loading, we may combine both expressions as follows:

$$V_{s1}^2/V_{s2}^2 = \sigma_1^{0.5}/\sigma_2^{0.5},$$

where V_{s1} and V_{s2} are the velocities and σ_1 and σ_2 are the effective stresses at times t_1 and t_2.

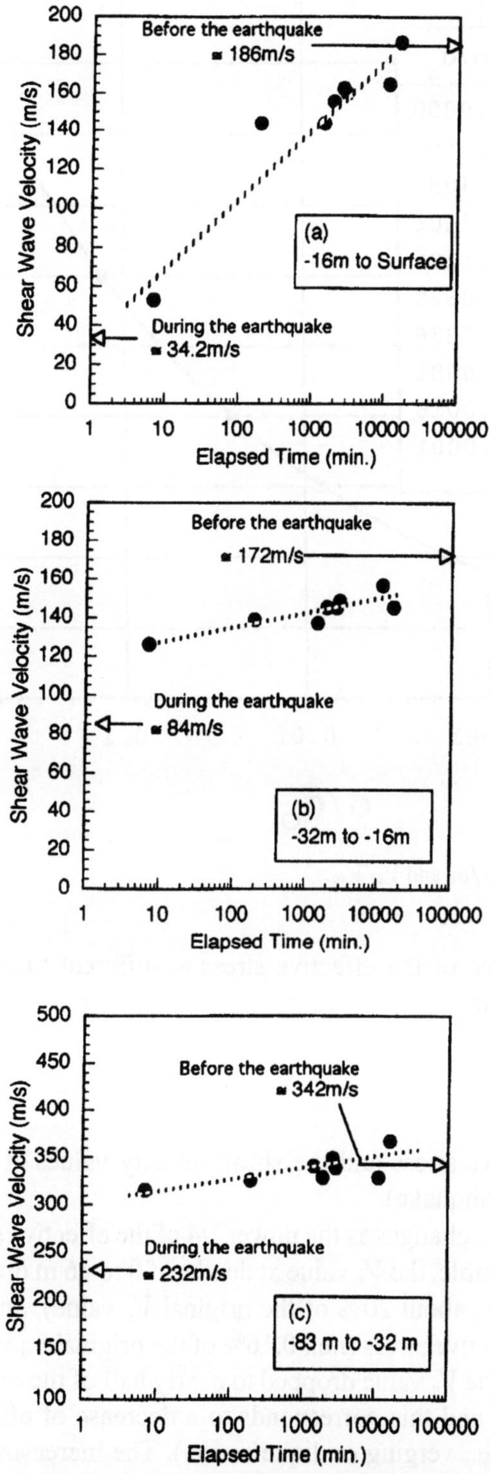

Figure 14. Increase of V_s with time at different depths.

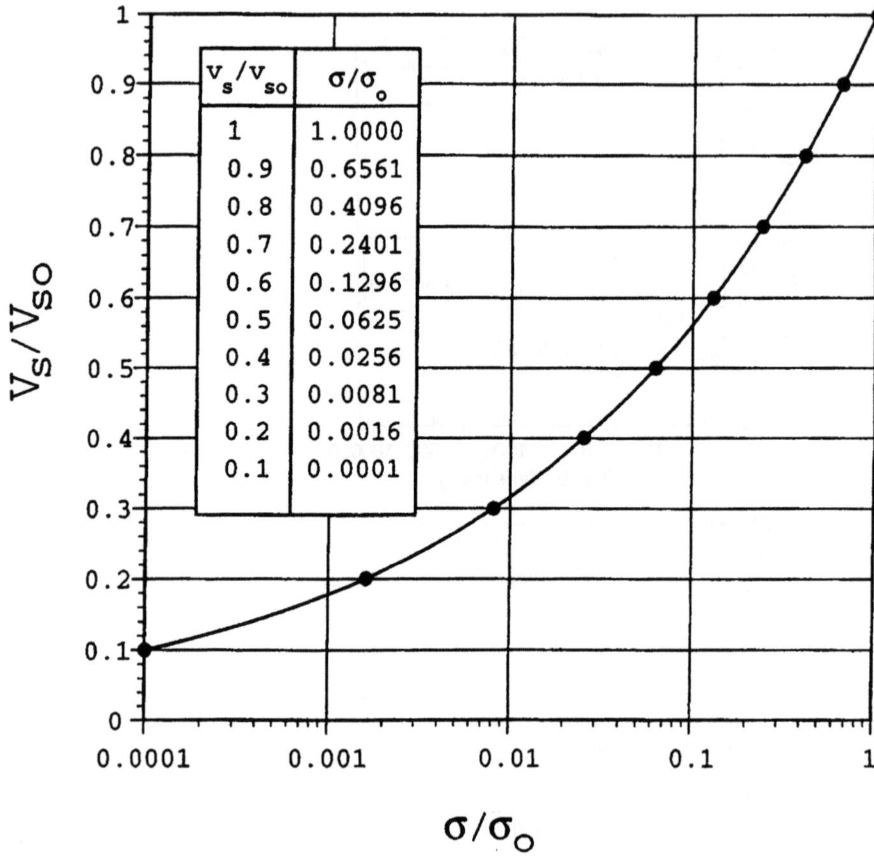

Figure 15. Relationship between σ/σ_0 and V_s/V_{s0}.

In order to evaluate the change of the effective stress at different times, the above equation may be modified as

$$(\sigma/\sigma_0) = (V_s/V_{s0})^4,$$

where σ_0 and V_{s0} are the effective stress and the shear velocity values at some reference time (i.e., before the earthquake).

This relationship implies that V_s changes as the power 1/4 of the effective stress, as indicated in Figure 15. For example, the V_s value at depths of 0 to 16 m dropped from 186 m s^{-1} to 34 m s^{-1} (i.e., about 20% of the original V_s value), and this means a near zero state of the effective stress (i.e., 0.16% of the original σ_0 value). Also at the depths of 16 to 32 m, the V_s value dropped to nearly half of the original value of 172 m s^{-1} to 84 m s^{-1} and this corresponds to a decrease of effective stress to 6.25% of the original (i.e., verging on liquefaction). The increasing rate of V_s at the depths of 16 to 32 m is slow, and after 10 days the state of effective stress is estimated to be still 50 to 60% of the original value.

5. Conclusions

Field data of seismic records and pore water pressure measurements are used to understand the time changes of ground stiffness after the strong motion of the Great Hanshin Earthquake. Because of the proximity of the epicentre to the man-made islands on soft seabed ground, the earthquake force caused a build-up of pore water pressure in the ground and catastrophic liquefaction in the superficial fill material in Port Island.

The pore water pressure measurement data as the seismic records of the main shock and of after-shocks give us a clear picture of how the excess pore pressure in the ground was built up and dissipated with time. Furthermore, the change of V_s during the earthquake may be used to estimate the change of effective stress and/or the build-up of excess pore water pressure due to the earthquake.

References

Iwasaki, Y., Kagawa, T., Sawada, S., Matsuyama, N., Ohshima, K., Ikawa, K., and Ohnishi, M.: 1994, Basement structure by air-gun reflection survey in Osaka Bay, southwest Japan, *Seismology* **46**(2), 395–403 (in Japanese)

Tanaka, Y.: 1996, The dynamic properties of marine clays as studied by in-situ testing, *Proc. 12th Southeast Asian Geotech. Conf.* **1**, 131–136

Hardin, B. O. and Richart, F. E.: 1963, Elastic wave velocities in granular soils, *Proc. of ASCE* **89**(SM1), 33–65.

Natural Hazards **00:** 287–296, 1997.

Mexico, San Francisco, Los Angeles and Kobe: What Next?

CINNA LOMNITZ
Institute of Geophysics, UNAM, 04510 Mexico, D.F.

(Received: 16 April 1997; in final form: 3 October 1997)

Abstract. Some analogies in the distribution of damage in the 1985 Mexico, 1989 Loma Prieta, and 1995 Kobe earthquakes may be attributable to similarities in the history of reclamation of bayshore or lakeshore environments by emplacing artificial fill on soft mud. In all three cases, a transitional environment has generated similar soil types and analogous forms of human settlement. These similarities may translate into hazardous situations because of amplification of seismic waves in wedge-shaped low-velocity layers; nonlinearity of seismic wave propagation in soft water-saturated soils; transitions from solid-like to liquid-like behavior, including liquefaction and the emergence of prograde surface waves; and other unforeseen conditions arising from surface geology. Severe stability problems may arise in tall, top-heavy structures and in structures with horizontal spans of the order of the wavelength of surface waves. Effective strategies of hazard reduction include a recognition of the many unanticipated ways in which earthquake hazard may become an emergent property of complex nature-society systems.

Key words: earthquake damage, San Francisco, Mexico City, Kobe.

1. Introduction

Patterns of settlement in urban areas are controlled by historical factors such as population growth, trade, and technological change. These factors are often responsible for the patchiness of the observed damage distribution in disasters. In the case of large recent earthquakes, damage has often been attributed to excessive ground motion or to the uneven or unanticipated performance of various types of structures. In this paper, we attempt to show, with specific examples to hand, how earthquakes may have an *archaeology* (Castaños and Lomnitz, 1995), which, when properly interpreted, may be recognized as a major factor of vulnerability. Such new concepts may assist in developing a consistent policy of preventing the recurrence of urban earthquake disasters.

Recent earthquake disasters in Mexico (1985), Loma Prieta (1989), Northridge (1994) and Kobe (1995) may be seen as marking a progression in severity. They are among the costliest natural disasters in history. The number of major natural disasters has been increasing by a factor of four since 1960 (Berz, 1995). Economic losses have risen at least by a factor of six in the same period, and the insured losses have been multiplied at least 14 times. The 1994 Northridge, California earthquake caused damage to the unprecedented amount of $30 billion, but this loss was topped the very next year by the Kobe, Japan earthquake. The annual loss of life in earthquakes, which had reached an all-time low between 1950 and 1960, is creeping up again and has now exceeded the levels of the 1920's, even when adjusting for population increase.

Such figures suggest that strategies of hazard reduction may have been less effective than was originally expected or desirable. The damage patterns have changed. Before 1950, most of the damage in earthquakes was directly attributable to traditional masonry construction. This fact was a powerful argument in support of earthquake engineering. After 1950, the incidence of this type of damage began to decrease and the number of failures in engineered structures increased, due to several poorly understood factors. Cities spread to areas on soft ground; costly high-rise buildings and fragile elevated multi-span structures were built on artificial fill and on soft natural soils which had been previously found unsuitable as foundation materials. Damage and loss of life were increasingly caused by the collapse of such structures and, equally significantly, of engineered structures built after 1945.

2. Damage Distribution in Kobe

Squeezed in a narrow strip of land between the Bay of Osaka and the Rokko Hills, Kobe could expand only in one direction: into the bay. Towards the end of the Edo period (1868), reclamation of Kobe Harbor had begun. After 1868 the city spread rapidly, and following near-total destruction in 1945 it was rebuilt in a hurry. Kobe Port became the main export facility of Japan and there were major pressures for urban expansion. Between 1966 and 1990, two large artificial islands, Rokko and Port, were reclaimed from the shallow bay by dumping *masado*, or crushed decomposed granite from the Rokko hills, directly on top of a layer of soft marine clay.

In the 1995 Kobe earthquake two elongated zones or 'belts' of damage were recognized (Figure 1). The upper belt, a few kilometers from the earthquake fault, was assigned the top intensity in the Japanese scale because it featured severe damage to two-storey homes, built after Kobe was razed in the 1945 air attacks. In the lower belt along the waterfront, however, damage to modern engineering structures was actually more impressive. Highways, bridges, elevated rail structures, port facilities and lifelines were destroyed. These heavily damaged structures had been carefully designed against horizontal earthquake forces. Several had spans in the order of 15–25 m, which apparently made them vulnerable to as yet unidentified surface waves of short wavelength.

The artificial soil known as masado was believed to be resistant to liquefaction; yet it liquefied on an unprecedented scale (Shibata *et al.*, 1996). Both damage belts ran along subsurface wedges of softer materials. The surface layers were not of uniform thickness, thus causing an observed patchiness of the damage pattern. For example, a conspicuous gap in the upper damage belt west of Sannomiya

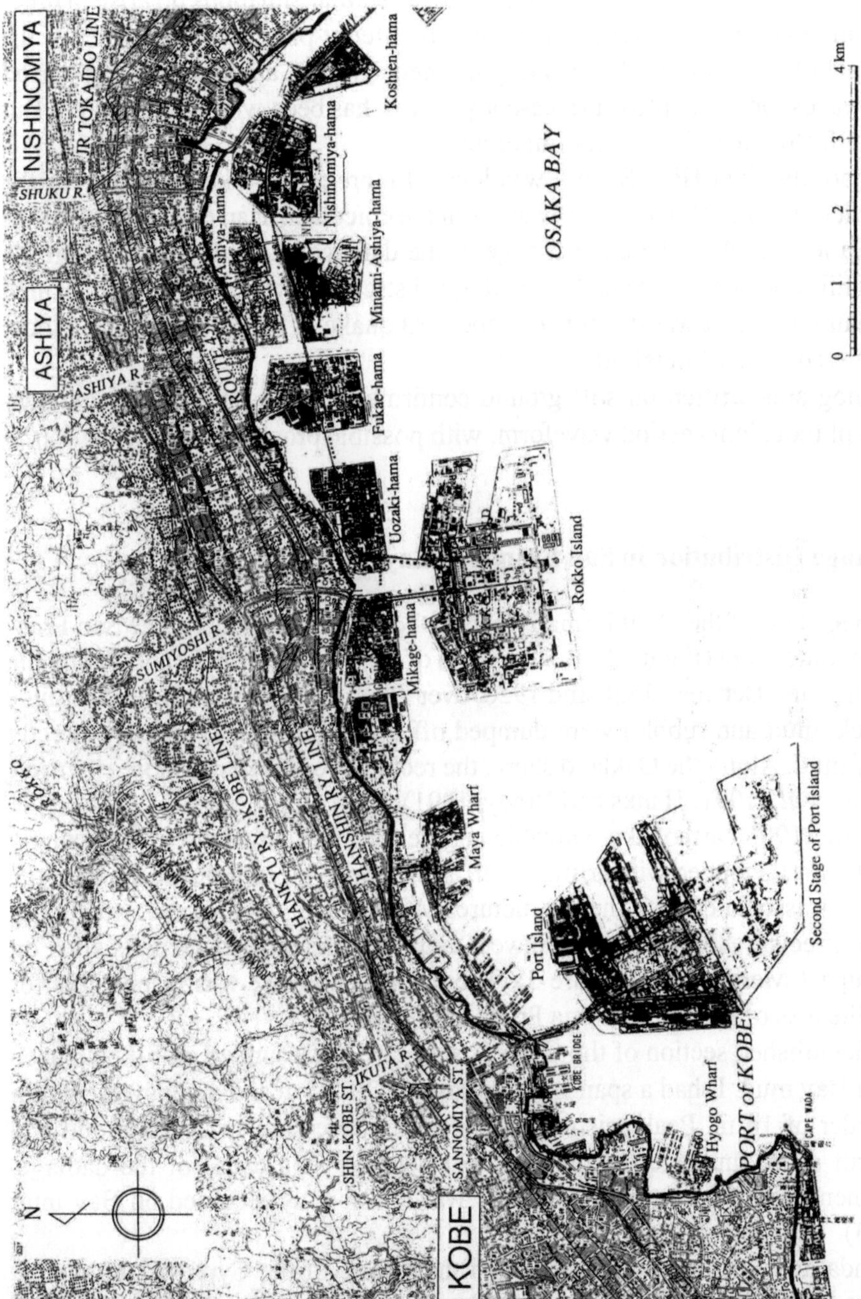

Figure 1. The two elongated zones ('belts') of the 1995 Kobe earthquake.

station, which had initially defied explanation (Kawase, 1996), may be due to an underground alluvial terrace which extends from the foothills toward the bay.

The broad parallelism of the damage belts, the waterfront, the tectonic structures and the pattern of human settlement suggests that the underlying processes must be causally connected, otherwise it would become necessary to appeal to coincidences (e.g., Kawase, 1996; Miyatake, 1996), as similar situations do arise in other waterfront situations. Subsurface geology admittedly plays a critical role, in the sense that soft shallow soil layers are prominent. In fact, an effect of long-period surface waves, or of coupled surface-body waves, has been widely postulated. But historical factors may be equally important.

Nishinomiya-kita High School was located more than 10 km inland from the main fault break and about 25 km away from the epicenter, at an elevation of 200 m above sea level and on the eastern edge of the damaged area. It was sited on the Ashiya Hills and not on the shallow geological structures which have been blamed for the damage. Yet it was destroyed in the earthquake. It had been built on a local pond covered with artificial fill.

Seismograms written on soft ground confirmed the presence of an unknown large-amplitude, long-period waveform, with possible prograde particle motion.

3. Damage Distribution in San Francisco Bay

The damage belt of the 1989 Loma Prieta earthquake extends along the San Francisco Bay waterfront (Figure 2). Reclamation of the bayshore was begun during the gold rush years. Between 1850 and 1920, over 20 million cubic yards of fill (dune sand, rock, mud and rubble) were dumped off the waterfront, most of it directly onto Bay mud. Along the Oakland shore, the reclamation was even more extensive (Chameau *et al.*, 1991; Hanks and Brady, 1991).

After the 1906 earthquake, valuable engineering structures were increasingly built on Bay mud topped with artificial fill, and in the 1989 earthquake much of the damage was associated with these structures. A section of the Cypress structure of Interstate Freeway 880 collapsed between adjacent undamaged sections sited on hard compact Merritt sand (Figure 3). The Cypress collapse was responsible for most of the loss of life in the Loma Prieta earthquake.

The demolished section of the Cypress structure was founded on fill emplaced on top of Bay mud. It had a span width of around 20–25 m and a transverse width of the order of 16 m. Peak seismic accelerations were similar for Bay mud and for Merritt sand; thus acceleration was not decisive as a cause of the damage. A prominent train of long-period surface waves was only recorded on Bay mud (Figure 4).

Miranda and Bertero (1991) attributed the failure of the Cypress structure to transverse vibrations at a frequency of 2.5 Hz, but Hanks and Brady (1991) pointed out that the spectrum of the signal peaked around 0.3 Hz, which is much too low

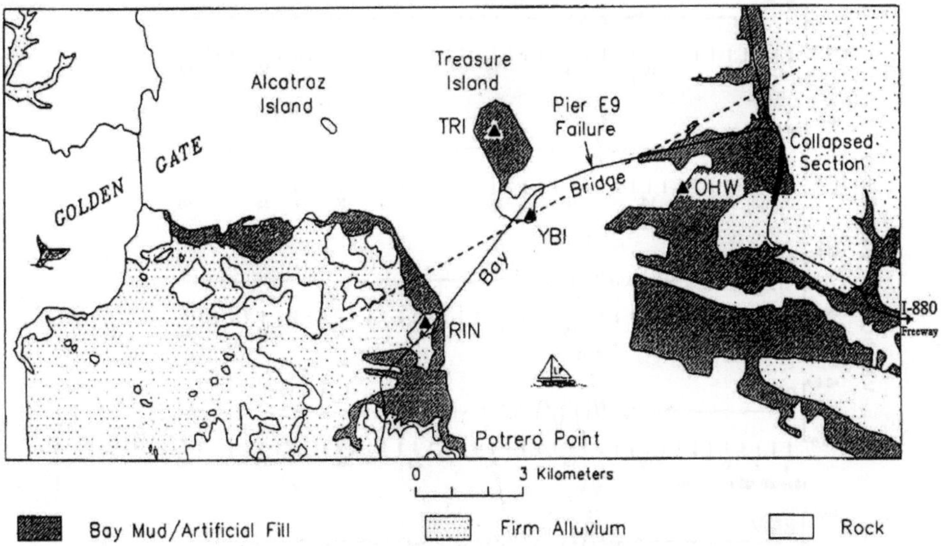

Figure 2. The damage belt of the 1989 Loma Prieta earthquake along the San Francisco Bay waterfront.

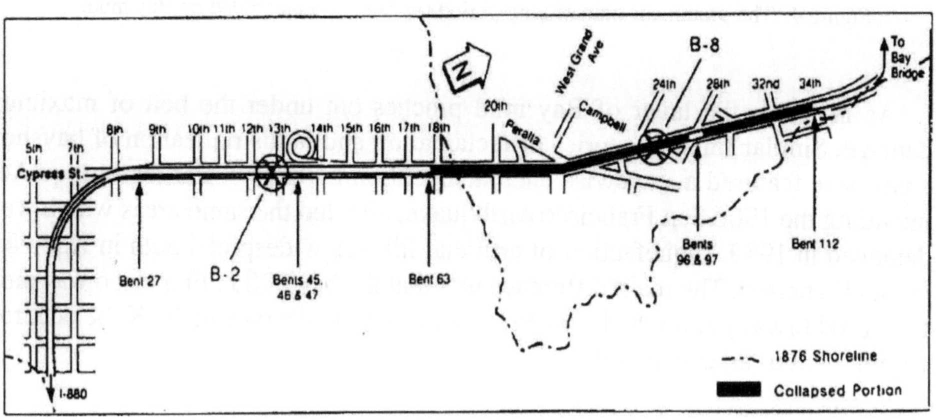

Figure 3. Detailed map of the Cypress Street viaduct section of I-880, showing the locations of the two borings to basement rock and the 1876 shoreline, outboard of which lie Bay muds and artificial fill (to the west in the *upper right*, to the south in the *lower left*). Modified from Housner (1990).

for such an explanation to be likely. Also, this theory failed to explain why the collapse should have occurred only in the section founded on Bay mud.

These discrepancies lend credibility to the suggestion that the collapse might have occurred by buckling in the axial mode, rather than by excessive transverse accelerations (Lomnitz, 1994). The observed periods of 1.5 s could have excited the structure efficiently by surface waves whose wavelength on Bay mud was in the order of the span of the structure.

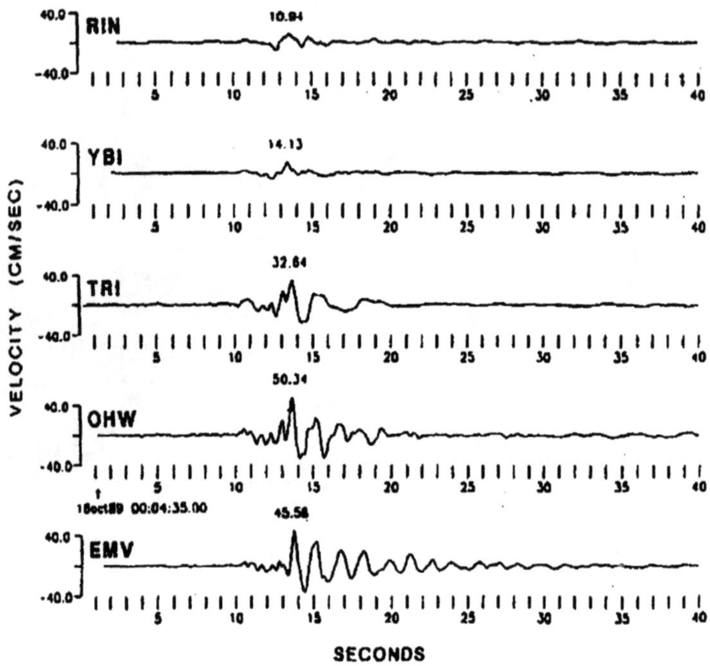

Figure 4. The prominent train of long-period surface waves recorded on Bay mud.

As in Kobe, the layer of Bay mud pinches out under the belt of maximum damage. Similar linked histories of reclamation and industrialization of bayshore areas have featured a stagewise encroachment on both bays. Earlier earthquakes, including the 1906 San Francisco earthquake, affected the same areas which were damaged in 1989. Liquefaction of artificial fill was widespread both in Kobe and in San Francisco. The main difference was that the bay of San Francisco was more than 100 km away from the Loma Prieta epicenter, while damage in Kobe occurred within 5 km of an active fault.

4. Damage Distribution in Mexico City

In the 1985 Mexico earthquake, the epicenter was about 400 km away from Mexico City; yet the damage distribution was quite similar to that in Kobe and San Francisco (Figure 5). Important structures collapsed on artificial fill emplaced on top of lake mud, and nowhere else. Areas on higher ground suffered only minor damage if any at all.

Mexico City was founded in 1325 by Aztecs who settled on a small island in a shallow lake. This lake occupied a flat intramontane depression of volcanic origin. In 1521, the city was destroyed by Spanish forces and the rubble or 'archaeological fill' was used as foundation material. The lake was drained in successive stages, mostly to prevent annual flooding. In the 19th century, the city began to expand

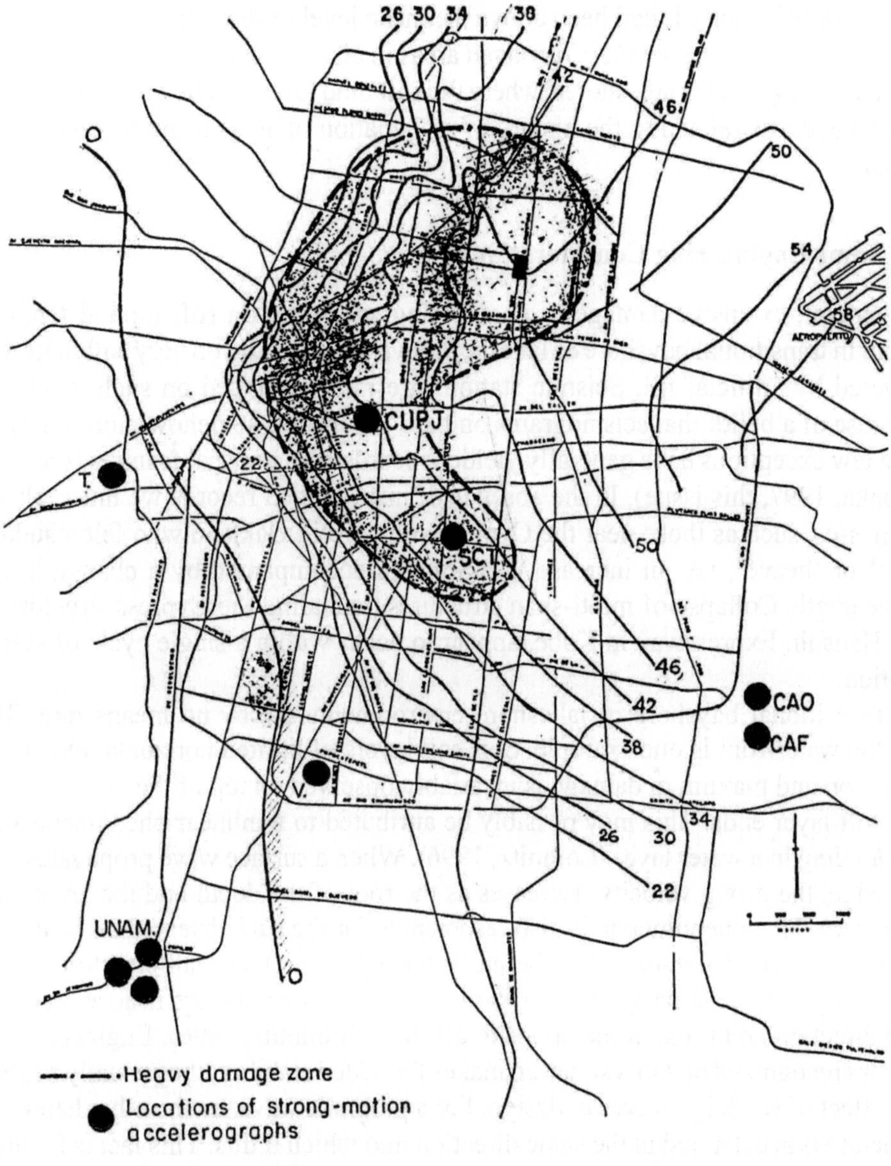

Figure 5. Plan of Mexico City showing locations of instrument stations and CUPJ site.

into the mudflats west of the island. Artificial fill was extensively used to improve foundation conditions in formerly waterlogged areas.

The damage distribution in the 1985 earthquake was similar to that in 1957 and other previous earthquakes. In detail, however, it was very patchy. Two main damage belts may be recognized: (a) the area running along the eastern edge of the former lake, (b) the area off the western edge of the former island. Between these two belts most recent urban development took place. The downtown area proper (i.e. the island where the city had been founded) suffered less severe damage, presumably because it had been above the mean level of the lake.

In conclusion, the severely damaged areas in all three earthquakes were located on mud along former lakeshores, where the soft mud layer pinched out. The pattern of damage was related to the history of reclamation of areas formerly covered by water.

5. Some Engineering Considerations

In general, extensive damage to engineering structures on soft ground tends to occur in transitional bayshore or lakeshore environments and on very soft materials covered by artificial fill. Seismic stations are rarely founded on such materials, because of a belief that seismograms on hard ground are somehow more relevant. The few exceptions have generally yielded records with unusual features (see, e.g., Tanaka, 1997, this issue). In the absence of instrumental records we must rely on witnesses, such as those near the Cypress structure in Oakland who felt a sudden 'jolt' or 'heave', i.e. an increase in amplitude accompanied by a change in the wavelength. Collapse of multi-span structures, including the Cypress structure or the Hanshin Expressway in Kobe, appear to occur within a single cycle of strong motion.

Transitional bayshore or lakeshore environments are by no means rare. The Seattle waterfront is one example. Soft soil layers of limited horizontal extent are common and maximum damage is invariably observed on top of the wedge where the soft layer ends. This may possibly be attributed to nonlinear phenomena akin to *shoaling* in a water layer (Lomnitz, 1996). When a surface wave propagates into a wedge, the group velocity decreases as the root of the depth and the amplitude increases. This phenomenon is also responsible for the surf observed on beaches.

Why should prograde waves be particularly hazardous to tall structures? In a retrograde surface wave (such as a Rayleigh wave), the structure is accelerated in one direction and tilted in the opposite direction (Lomnitz, 1996). Engineers have always recognized that this situation makes for added stability; they usually neglect the effect of Rayleigh waves in design. For a prograde wave, on the other hand, the structure is accelerated in the same direction into which it tilts. This fact is familiar to naval engineers, who must deal with the effects of prograde water waves on naval structures. This is the main reason why ships are designed with a low center of mass.

6. Discussion and Conclusions

The peak horizontal ground acceleration in the 1985 Mexico earthquake was only 0.2 g on soft ground, yet the estimated surface strains were very large, because of the long duration of strong ground motion and the severe degradation of the shear modulus of the mud (Lomnitz, 1996). In transitional bayshore and lakeshore environments the surface shear-wave velocity is often below 100 m s^{-1}. Thus the wavelength at periods of 1–2 s can be in the order of the length of a building, or of the span between piers of bridges, or between bents of viaducts.

Buckling of vertical structural members is frequently observed on soft ground (Nonaka *et al.*, 1995). Such failures are typically caused by large rotations and by spatial resonance. Thus, if the wavelength of a surface wave approaches the major horizontal dimensions or the span width of a structure, adjacent vertical structural members may rotate in a synchronous mode. The resulting harmonic resonant motion can grow to very large amplitudes and is not normally considered in structural design. For example, in the case of the Cypress structure, buckling of the upper roadway produced such severe tensional stresses that the upper-level vertical columns were torn out from their bases and ejected vertically into the air. In Kobe, steel silos were buckled and shortened in a vertical direction as a result of systematic rotation about a vertical axis.

Such severe dynamic effects arise only on very soft ground. They must be attributed to the presence of very large rotational strains. The striking analogies between the history and the configuration of surface geology and damage in Kobe, San Francisco and Mexico City suggest that near-surface geology might play a greater role in damage causation than has been generally acknowledged. The history of human settlement, especially in bayshore and lake environments, may be a major unrecognized factor of earthquake hazard.

Disasters have an archaeology in the sense that human decisions (e.g., involving the use of manmade fill to extend urban areas into a bay or a lake) can have long-lasting consequences in terms of the generation of future disasters.

In conclusion, disasters may be considered as emergent properties of complex nature-society systems. The earthquake hazard is not static: it evolves with the system. Sadly, we always seem to be ready for the previous earthquake and hardly ever for the next one.

Acknowledgement

This note was suggested by an oral comment made by Raul Madariaga after the presentation of my paper at the 1996 ESC Conference in Reykjavik, Iceland.

References

Beresnev, I. A. and Wen, K.-L.: 1996, The possibility of observing nonlinear path effects in earthquake-induced seismic wave propagation, *Bull. Seismol. Soc. Am.* **86**, 1028–1041.
Berz, G.: 1995, Major disasters in 1994, Munich Reinsurance Corp., Munich.
Biot, M.: 1940, The influence of initial stress on elastic waves, *J. Appl. Phys.* **11**, 522–530.
Casagrande, A.: 1936. Characteristics of cohesionless soils affecting the stability of slopes and earth fills, *J. Boston Soc. Civ. Eng.*, January.

Chameau, J. L., Clough, G .W., Reyna, F., and Frost, J. D.: 1991, Liquefaction response of San Francisco bayshore fills, *Bull. Seismol. Soc. Am.* **81**, 1998–2018.
Darwin, C.: 1846, *Geological Observations on South America*, London.
Gilbert, F.: 1967, Gravitationally perturbed Rayleigh waves, *Bull. Seismol. Soc. Am.* **57**, 783–794.
Gilbert, F., Laster, S. J., Backus, M. M., and Schell, R.: 1962, Observation of pulses on an interface, *Bull. Seismol. Soc. Am.* **52**, 847–868.
Hanks, T. C. and Brady, A. G.: 1991, The Loma Prieta earthquake, ground motion, and damage in Oakland, Treasure Island, and San Francisco, *Bull. Seismol. Soc. Am.* **81**, 2019–2047.
Hardin, B. O. and Drnevich, V. P.: 1972, Shear modulus and damping in soils: Design equations and curves, *J. Soil Mech. Found. Div., Proc. Am. Soc. Civil Eng.*, SM7, **98**, 667–692.
Infeld, E. and Rowlands, G.: 1990, *Nonlinear Waves, Solitons and Chaos*, Cambridge University Press.
Korteweg, D. J. and de Vries, G.: 1895, On the change of form of long waves advancing in a rectangular canal and on a new type of long stationary waves, *Philos. Mag.* **39**, 422–443.
Lomnitz, C.: 1996, The gravilastic equation and the emergence of gravity waves in large earthquakes, *Bull. Seismol. Soc. Am.* **86**, 1220–1228.
Miranda, E. and Bertero, V. V.: 1991, Evaluation of the failure of the Cypress viaduct in the Loma Prieta earthquake, *Bull. Seismol. Soc. Am.* **81**, 2070–2085.
Miyatake, T.: 1996, 3-D simulation of near-field strong motion based on dynamic modelling: basin edge effect derived from rupture directivity, *Geophys. Res. Lett.*, submitted.
Nonaka, T., Mimura, K., and Tanimura, S.: 1995. A note on structural failure due to the Great Hanshin Earthquake, *Materials Sci. Res. Int.* **1**, 132–133.
Shibata, T., Oka, F., and Ozawa, Y.: 1996, Characteristics of ground deformation due to liquefaction, *Soils and Foundations (Japan)*, Special Issue, January, 65–79.
Tanaka, Y.: 1997, The behaviour of a man-made island during the Great Hanshin Earthquake, Japan, *Natural Hazards* **16**, 267–285.

Natural Hazards **16:** 297–300, 1997.

Meeting Report:

Hazards-96: Concluding Plenary Session and Discussion

MADHAV L. KHANDEKAR
Atmospheric Environment Service, Climate Research Branch, Downsview, Ontario, Canada

(Received: 31 July 1997)

Hazards-96 (Sixth International Symposium on Natural and Man-Made Hazards, Toronto, Canada, 21–26 July 1996) brought together scientists, engineers and government officials from over 15 countries to learn more about major natural disasters of the nineties and their socio-economic impact at local/regional, national and international levels. Over one hundred papers on scientific, technical and socio-economic aspects of natural disasters were presented.

The objective of the concluding plenary session was to capture the main theme of the conference and to assess our knowledge towards mitigating future hazards. In order to facilitate discussion, the plenary session opened with presentations by four experts, G. Papadapoulos (Geodynamic Institute, National Observatory of Athens, Greece), J. Pudykiewicz (Canadian Meteorological Centre, Montreal, Canada), G. Watson (Disaster Prevention Research Institute, Kyoto University, Japan) and Barbara Siegal (New Mexico State University, Las Cruces, U.S.A.). The first three speakers provided an overview of papers presented at the symposium on the specific area of their expertise and identified research needs and various measures of mitigation like public education, warning systems and international co-operation. The last speaker (Barbara Siegal) spoke on various mitigation strategies and assessed areas of strengths and weaknesses in the mitigation measures. Several important points were brought out during the experts' presentations and these points are summarized as follows:

Seismic Hazards

Three aspects of seismic hazards were brought out by Papadapoulos during his presentation:

1. Need to determine and standardize type and quality of data.
2. Need for a stochastic model and a re-examination of the classical model which uses the G-R relationship.

3 Mitigation strategies based on lessons learned from previous and recent seismic hazards (ex. behaviour of particular structures or structural elements like pipelines, bridges and highways).

Papadapoulos proposed in his presentation a need for 'seismic scenarios' for developing appropriate response to earthquakes in megacities.

Atmospheric Hazards

In his presentation, Pudykiewicz discussed the various scales of atmospheric motion and pointed out the difficulty in predicting very small scales like a tornado or very large scales like climate change on a global scale. The synoptic scale of atmospheric motion which represent the day-to-day weather movement can be reasonably well predicted at present. The impact of chemical hazards on atmosphere like increased greenhouse gas concentration due to burning of fossil fuels or volcanic eruptions is reasonably well understood but is not well predicted yet because of complexity and interdependence of atmospheric motion on various scales. Pudykiewicz further discussed the impact of Kuwait oil fires (March 1991) on local and regional weather with a short-lived climatic impact, while the eruption of Mount Pinatubo (June 1991) producing a much longer-lived climatic impact which could be identified for up to two years after the eruption.

Present numerical weather prediction models have achieved a level of skill which can provide mitigation of weather related hazards on a time scale from a few hours to a few days.

Marine Hazards

Several types of marine hazards were surveyed and briefly discussed by Watson. Among the hazards discussed were, the interannual variation of hurricane frequency, tsunami hazards for Greece and in the Aegean Sea, storm surges off the coasts of Australia, Japan and in the Bay of Bengal and cyclone disasters in Bangladesh. The impact of recent marine hazards were highlighted, in particular the 1991 Bangladesh cyclone (which resulted in over one hundred thousand deaths) and the hurricane Andrew in 1994 (which inflicted massive property damage along the coast of Florida, U.S.A.). Watson further summarized the lessons to be learnt from recent marine hazards as

1. increasing risk due to rapid development of coastal regions
2. identifying the risk zones through research and
3. developing adequate education and communication skill for public as well as for industry (ex. advice to insurance industry).

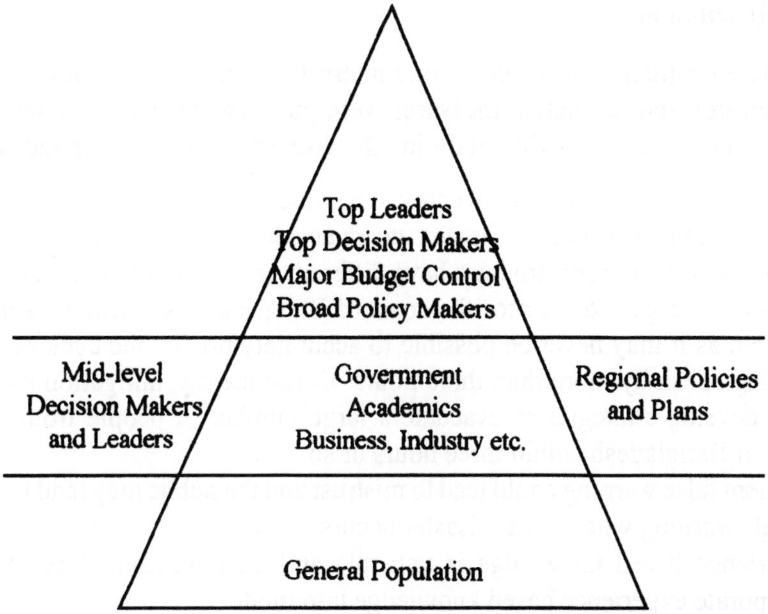

Figure 1. 'Top-down' structure of government framework for mitigation measures.

Mitigation

Siegal provided a comprehensive overview of various types of mitigation and development of mitigation strategies. She discussed two types of mitigation measures namely active (ex. preparation) and passive (ex. law enforcement, insurance coverage etc.). The various mitigation strategies and the societal responses were discussed through case studies and examples. She further described the conventional 'Top-down' structure of government framework for mitigation measures (see Figure 1) as ineffective, since the general population will not be told what to do in case of a natural (or a technological) disaster. She concluded her presentation by identifying seven mitigation measures applicable to public as well as decision makers:

1. Learn from experience.
2. Acknowledge human error.
3. Improve communication.
4. Increase 'technology'.
5. Become more interdisciplinary.
6. Err on the side of safety and
7. Buy insurance.

General Discussion

During the open floor discussion, several interesting points on a variety of topics relating hazards and mitigation measures were put forward and discussed. Some of the important issues brought out during the discussion are summarized below:

1. Rating of a disaster using conventional parameters like the number of deaths, property damage, economic impact etc. does not necessarily work everywhere as human life and infrastructure have different values in different countries.
2. The storm surge problem for the coast of Bangladesh is a difficult scientific problem as it may never be possible to accurately predict the cyclone arrival in Bangladesh by more than three hours. Consequently, mitigation measures must develop strategies to evacuate a large number of people from coastal areas of Bangladesh within three hours or so.
3. Frequent false warning could lead to mistrust and the public may tend to ignore a 'real' warning when a real disaster occurs.
4. Experience based knowledge is valuable and attempts should be made to incorporate experience based knowledge into models.
5. Scientists (and researchers) can communicate well with each other but not so well with the general public. There is a need to provide a basic education to the public about various types of hazards and how they occur and how the society can prepare for it.
6. It is not possible to prevent all disasters. There is a need to identify permissible hazards or preventable/mitigable disasters or acceptable/mitigable disasters.

Acknowledgements

It is a pleasure to acknowledge the help of Ms. Kim Parkhill of the Air Quality Research Branch, Atmospheric Environment Service (AES) for taking notes during the plenary session and preparing a summary. Thanks are also due to Ms. Eileen Jeffers of the Climate Research Branch, AES for the excellent typing of this report.

MIX
Papier aus verantwortungsvollen Quellen
Paper from responsible sources
FSC® C105338

If you have any concerns about our products,
you can contact us on
ProductSafety@springernature.com

In case Publisher is established outside the EU,
the EU authorized representative is:
Springer Nature Customer Service Center GmbH
Europaplatz 3, 69115 Heidelberg, Germany

Printed by Libri Plureos GmbH
in Hamburg, Germany